CatchUp

Chemistry

for the life and medical sciences

CatchUp

Chemistry 2nd Edition

for the life and medical sciences

Mitch Fry

Faculty of Biological Sciences, University of Leeds, UK

and

Elizabeth Page

School of Chemistry, University of Reading, UK

Scion

Second edition published 2012

First edition published 2005, and reprinted 2006, 2007, 2008, 2009, 2010, 2011

A CIP catalogue record for this book is available from the British Library.

ISBN 978 1 904842 89 7

Scion Publishing Limited
The Old Hayloft, Vantage Business Park, Bloxham Rd, Banbury, OX16 9UX, UK
www.scionpublishing.com

Important Note from the Publisher

The information contained within this book was obtained by Scion Publishing Limited from sources believed by us to be reliable. However, while every effort has been made to ensure its accuracy, no responsibility for loss or injury whatsoever occasioned to any person acting or refraining from action as a result of information contained herein can be accepted by the authors or publishers.

Typeset by Phoenix Photosetting, Chatham, Kent, UK
Printed by 4edge Limited, Hockley, UK

Contents

Preface to second edition

The first edition of the book proved to be very popular, but we received some useful suggestions as to other areas that bioscience students need to understand. So this new edition contains new sections on the gas laws, macromolecules, biological transport mechanisms and transporters, reaction mechanisms and reaction kinetics, and the section on functional groups has been significantly expanded.

We hope that this new edition continues to cover the chemistry required in most life and medical science courses.

Mitch Fry and Elizabeth Page
April 2012

Preface to the first edition

Life is a collection of interacting chemical processes, designed to be self-sustaining. The carefully managed oxidation and release of energy from foodstuffs allows life-forms to maintain a precisely controlled cellular environment, and to build ever increasing levels of metabolic and structural complexity. All life on earth is built upon basic chemical principles; in particular it has taken the element carbon as a building block for producing a huge and diverse range of biological molecules, all designed to function within the universal solvent of life, water.

Nowadays, in our rush to achieve academically, there is a tendency to 'package' information. While this may 'fast-track' our progress, it actually dilutes our understanding. Seeing the whole picture is so much more rewarding. Studying the basic chemical principles of life is necessarily the first step in developing that picture.

You don't need a degree in chemistry to be aware of these principles. This short textbook explores those chemical principles which are readily extrapolated to the biological scenario, providing a basic foundation to facilitate an understanding of biological processes. We have developed these principles in an easy to understand way, without assuming any prior knowledge of chemistry beyond that covered in secondary school. You cannot divorce biology from chemistry; the former is a special extrapolation of the latter. Nor should you avoid taking the time to understand 'first principles'; your effort will be amply rewarded.

As you embark upon your chosen bioscience course, take the time to wonder at the marvel that is life; ask why? and how? Engage your enthusiasm; it's what being a biological scientist is all about!

Mitch Fry and Elizabeth Page
Leeds and Reading, May 2005

About the authors

Dr Mitch Fry BSc PGCE PhD is a biochemistry graduate who has worked as both a senior research scientist in the pharmaceutical industry and as a science teacher in secondary and higher education. This has included the teaching, supervision and support of life science and medical undergraduate students, at the University of Leeds and Sheffield Hallam University, including 'pre-university' awareness activities and university admission procedures.

Dr Elizabeth Page BSc PhD PGCE is Senior Lecturer in Chemistry Education and Director of Undergraduate Studies in the School of Chemistry at the University of Reading. She has had experience in teaching chemistry to biologists and other life sciences students for the past fifteen years. She has particular interest in supporting first year undergraduates as they make the transition to tertiary education.

01 Elements, atoms and electrons

BASIC CONCEPTS:

We begin by considering basic atomic structure and the nature of isotopes. Isotopes play an important role in biology and are the subject of a 'Taking it Further' section. We look at electron distribution and configuration in atoms and explore the concept of atomic orbitals. An appreciation of atomic orbitals is essential to an understanding of the reactivity and bonding behaviour of atoms, and in particular to those elements that constitute the major building blocks of biological systems.

The ninety-two naturally occurring elements can combine in a variety of ways to form the matter that constitutes the world we live in. An **element** is a single substance which cannot be split by chemical means into anything simpler; for example, carbon is an element, oxygen is an element. Every element is represented by a symbol, an upper case letter, followed in some cases by a lower case letter; for example, carbon = **C**, calcium = **Ca**, nitrogen = **N**, sodium = **Na**. Elements are made up of many tiny, but identical particles, called **atoms**. An atom can be described as the smallest particle into which an element can be divided, while still retaining the properties of that element. Thus the element carbon is composed entirely of carbon atoms, the element oxygen is composed entirely of oxygen atoms, and so on. Atoms themselves consist of smaller units known as **sub-atomic particles**. There are three main types of sub-atomic particles, known as **protons, neutrons** and **electrons**.

The relatively massive centre of an atom is referred to as the **atomic nucleus**. The atomic nucleus is composed of protons and neutrons. These two sub-atomic particles are distinguished by their charge; each proton carries a single positive charge, whereas neutrons have no net charge (hence their name). The number of protons present in the nucleus is unique to each element. The **atomic number (*Z*)** of an element is equal to the number of protons in its nucleus. The sum of the number of protons and neutrons in the nucleus of an atom is equal to the **mass number (*A*)** of that element.

REMINDER

Atomic number (Z) = number of protons
Mass number (A) = number of protons + number of neutrons

1.1 Isotopes

Sometimes the number of neutrons in an element's nucleus can vary, giving rise to different **isotopes**.

The nuclear composition of an atom is shown by $^{A}_{Z}\mathbf{X}$ where A = the mass number and Z = the atomic number.

Because the atomic number of an element is specific to that element this symbol for the atom is often reduced to ^{A}X.

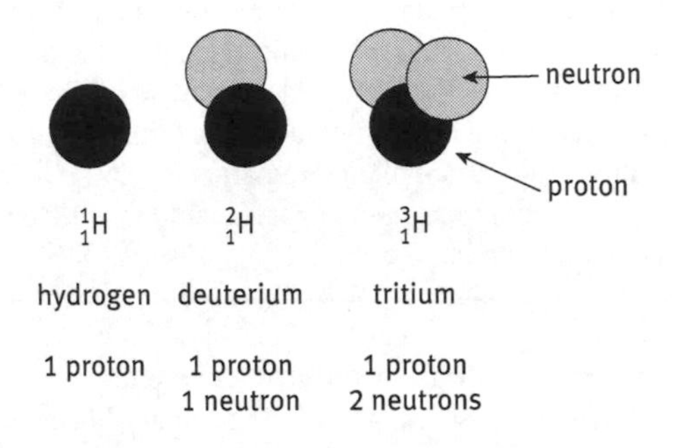

Figure 1. Isotopes of hydrogen

Hydrogen may exist as $^{1}_{1}$H (also written hydrogen-1), $^{2}_{1}$H (also written hydrogen-2) or $^{3}_{1}$H (also written hydrogen-3). In other words, $^{1}_{1}$H (hydrogen, the commonest form) has 1 proton only in its nucleus and has a mass number of 1. $^{2}_{1}$H (referred to as deuterium) has 1 proton plus 1 neutron in its nucleus, giving it a mass number of 2. $^{3}_{1}$H (referred to as tritium) has 1 proton plus 2 neutrons, giving it a mass number of 3. In each case there is only one proton present, and so the element is hydrogen; these are all isotopic forms of hydrogen (Fig. 1). Likewise, carbon can exist as three different isotopes, $^{12}_{6}$C, $^{13}_{6}$C, and $^{14}_{6}$C of which $^{12}_{6}$C and $^{14}_{6}$C are the most common. Each of these is an isotope of carbon, each contains 6 protons, but the number of neutrons varies from 6 to 8.

REMINDER

Isotopes of the same element differ from each other in their number of neutrons, not in their number of protons

Isotopes may be either **stable** or **radioactive**. In the examples we have used, $^{1}_{1}$H and $^{2}_{1}$H are stable isotopes of hydrogen, whereas $^{3}_{1}$H is radioactive. Similarly, $^{12}_{6}$C and $^{13}_{6}$C are stable isotopes of carbon, but $^{14}_{6}$C is radioactive.

TAKING IT FURTHER: Isotopes in biology (p. 9)

Isotopes are extremely useful to biologists and are used in a variety of applications.

REMINDER

Radioisotopes are unstable and decay, releasing radioactive emissions

1.2 Electrons

The number of electrons in an atom equals the number of protons in the nucleus of that atom. It is the arrangement of electrons in an atom which determines its chemical reactivity.

Electrons can be thought of as sub-atomic particles having almost negligible mass. The single negative charge (–1) on an electron is equal and opposite to the positive charge on the proton. The number of electrons in an atom balances the number of protons, so an atom has no overall net charge.

Electrons move around the nucleus of the atom with speeds close to that of the speed of light. It is almost impossible to say exactly where an electron is at any specific point in time; this is the basis of the uncertainty principle and we talk about 'the probability of finding an electron in a particular space at any point in time'.

Experiments carried out at the beginning of the twentieth century demonstrated that electrons do not freely circulate around the nucleus but are restrained to specific **energy levels** or **shells**. Energy levels are given numbers, $\boldsymbol{n}$, starting with $n = 1$; under normal conditions electrons occupy the lowest energy levels first. (*The terms 'energy level' and 'shell' tend to be used interchangeably.*)

REMINDER

While the atomic number of an atom defines the element, it is the arrangement of electrons that determines the chemical reactivity

Within each energy level are sub-levels, which are specific locations within an energy level where there is a high probability of finding an electron. These regions within an energy level are referred to as **atomic orbitals.** Atomic orbitals have particular shapes and can hold a maximum of two electrons. Closest to the nucleus is the lowest energy level with the value $n = 1$. The particular type of orbital within the $n = 1$ level is called an **s** orbital. Because it is in the $n = 1$ energy level it is called the **1s** orbital. The s orbital is spherical in shape.

REMINDER

Atomic orbitals are regions in space where there is a high probability of finding electrons

We can define the **1s** orbital as a spherical region of space close to and surrounding the nucleus where there is a high probability of finding an electron (Fig. 2). The small black spot at the centre of each diagram represents the nucleus of the atom. The diagram is not to scale. Most of the space outside the nucleus of the atom is empty. In fact the radius of the nucleus is roughly one ten-thousandth of the size of the whole atom.

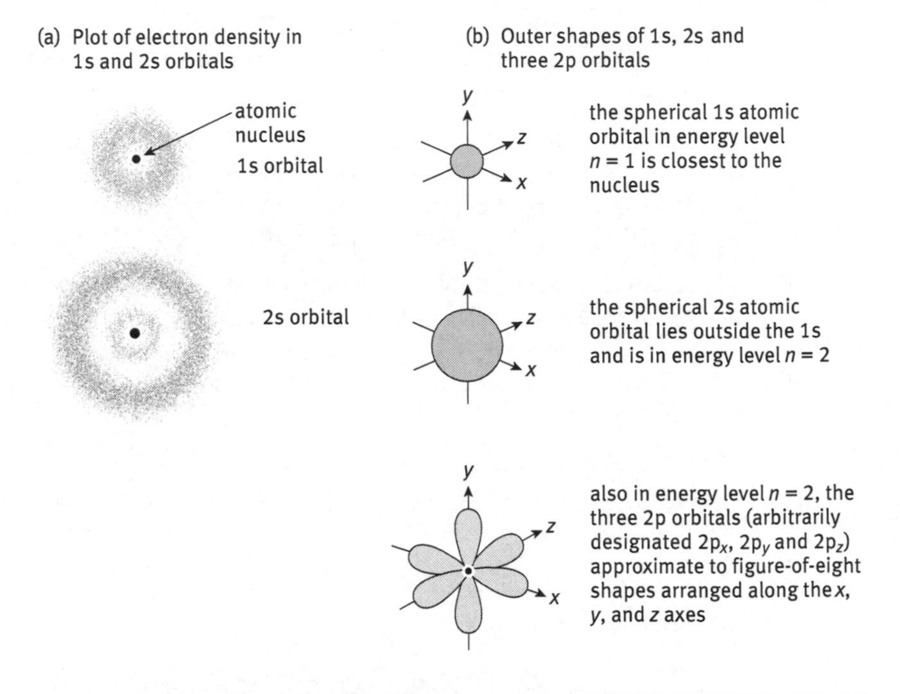

Figure 2. Arrangement of *s* and *p* atomic orbitals

The second energy level ($n = 2$) is at a slightly higher energy than the first. Within the second energy level there are two types of regions in space where an electron can be found. One of these regions is a spherical orbital, as in the first energy level. However, the name of this orbital is the 2s orbital because it is in the second energy level.

The second type of atomic orbital at this energy level (n = 2) is called a **p** orbital. There are three p orbitals within an energy level, and each p orbital has a specific orientation along the *x*, *y* or *z* axis. p orbitals represent regions in space which are shaped like a figure of eight or a dumb-bell. The labels given to the three p orbitals within the second level are $\mathbf{2p_x}$, $\mathbf{2p_y}$ and $\mathbf{2p_z}$. As we move further from the nucleus the levels become increasingly higher in energy and the orbitals within them become more complex and numerous.

There are some important rules governing the way electrons fill energy levels and orbitals.

- **Electrons fill orbitals of the lowest energy first.** So the first level ($n = 1$) with its 1s orbital is filled before an electron can occupy the second level.
- **Within an energy level electrons fill the orbitals with the lowest energy first.** s orbitals represent a lower energy configuration than p orbitals and so electrons will fill the s orbital within an energy level before filling the p orbitals.
- **Individual orbitals can hold a maximum of two electrons.** Each s orbital (1s, 2s, 3s etc.) can only accommodate two electrons at the most. When two electrons occupy the same orbital they spin in opposite directions, otherwise their negative charges would force them apart.

It is rather difficult to draw electron orbitals but much easier to think of the way in which electrons fill orbitals as **'electrons in boxes'**. This is usually referred to as the **electron configuration** of the atom.

For example, hydrogen is the simplest atom with only one proton in the nucleus and therefore one electron outside the nucleus. We can write the electron configuration of hydrogen as:

$$1s^1$$

This means that the single electron of the hydrogen atom occupies the 1s orbital, which is in the lowest energy level, $n = 1$. The next element is made by adding one more proton to the nucleus and one more electron to the outer energy level. This element is helium, He, and it has two neutrons in its nucleus. The symbol for helium is $^{4}_{2}He$. The two electrons of the helium atom occupy the 1s orbital and so the electron configuration of helium is:

$$1s^2$$

The first energy level is now full and the next element lithium, Li with an atomic number of 3, has three electrons. Two of the electrons from lithium occupy the first level and the remaining electron must go into the s orbital of the second level, $n = 2$. The electron configuration of lithium is therefore:

$$1s^22s^1$$

When the 2s orbital is full the electrons start to fill the 2p orbitals, which are of slightly higher energy (but still in energy level $n = 2$). So, by the time we reach carbon, which has 6 electrons, the electron configuration can be written as:

$$1s^22s^22p^2$$

The three 2p orbitals ($2p_x$, $2p_y$, $2p_z$) are considered to have identical energies. Therefore, an electron will enter an empty p orbital before pairing with another electron in a half-full p orbital. As all the 2p orbitals are equivalent in energy it is not possible to say which of the 2p orbitals ($2p_x$, $2p_y$ or $2p_z$) is filled first. As there are three p orbitals in each level (except $n = 1$) and each can hold a maximum of two electrons, a total of six electrons can occupy the p orbitals in any energy level.

The information above can be summarised pictorially by using the 'electrons in boxes' representation. This is shown here for carbon, C (Fig. 3).

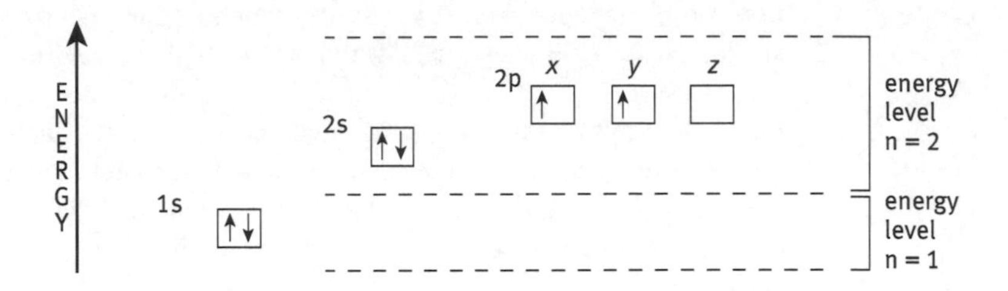

Figure 3. Electron filling in carbon

In energy level $n = 1$, the 1s orbital is full, containing two electrons of opposite spin direction, shown by the two arrows pointing in opposite directions. Likewise, at a higher energy level ($n = 2$), the 2s orbital is full, and at a slightly higher energy still, but within the second level, there is one electron in two of the three p orbitals and one p orbital is empty. Here the electrons have been arbitrarily assigned to the $2p_x$ and $2p_y$ orbitals.

Each element in the periodic table is formed by placing one more proton in the nucleus and one electron in the outer energy levels to balance the charge. After carbon, with atomic number of 6, is nitrogen (atomic number 7); the electron configuration of nitrogen is shown in Fig. 4.

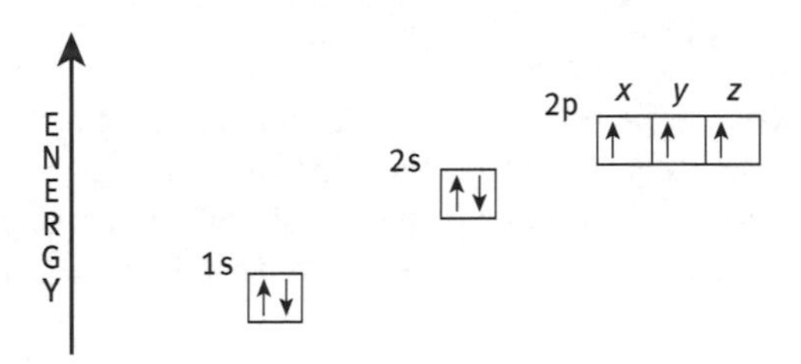

Figure 4. Electron filling in nitrogen

In the nitrogen atom each p orbital contains one electron and the 2p orbitals are half-filled.

Introducing one more proton in the nucleus and one electron in the outer energy level gives the element oxygen with atomic number 8. The extra electron of oxygen must now enter a half-filled orbital and so adopt an opposite spin. Thus one of the 2p orbitals is now full (Fig. 5).

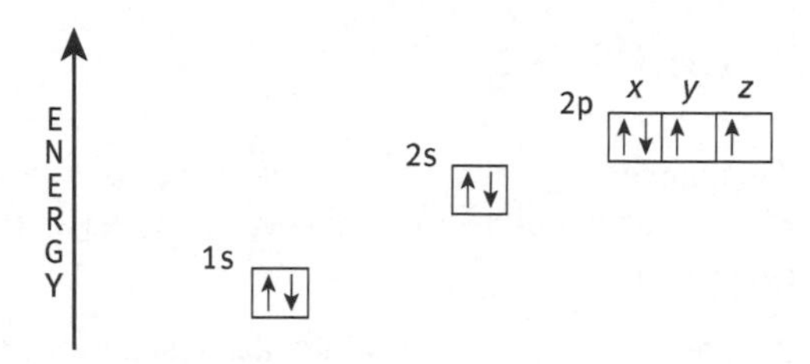

Figure 5. Electron filling in oxygen

Fluorine (atomic number 9) has two electrons in two of the 2p orbitals and one electron in the third 2p orbital. By the time we reach neon (atomic number 10), at the end of the second row in the periodic table, all the orbitals in the second level are full.

Neon is an unreactive element and a member of the inert or noble gas group. Its lack of reactivity resides in the fact that all its outer electron orbitals are full. This means that neon is reluctant to either gain or lose electrons to other atoms and so does not easily take part in bonding. The electronic configuration of neon is shown in Fig. 6.

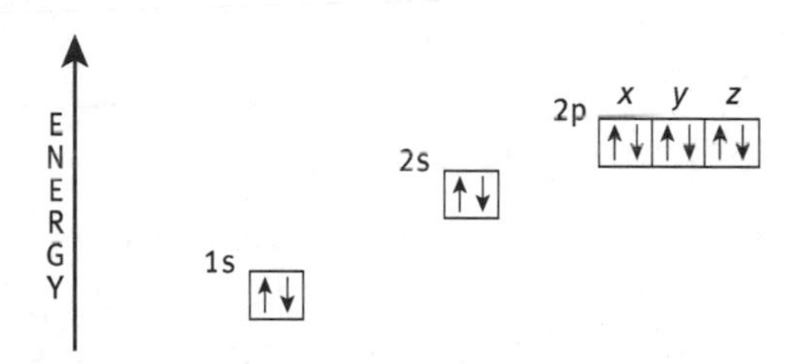

Figure 6. Electron filling in neon

TAKING IT FURTHER:
The periodic table
(p. 13)

It is simplistic, but useful, to think of atoms reacting together in an attempt to fill their outer electron energy levels and in doing so to reach a stable state (the **octet rule**). This model can be used to describe the way elements in the second row of the periodic table undergo bonding. Elements in the group which is furthest to the right of the periodic table, such as helium, neon, argon and krypton, all have filled outer energy levels and are very stable and unreactive. Reactivity, which results in the formation of new bonds between atoms, is most often achieved through the 'sharing' of electrons in the formation of covalent bonds. By sharing electrons, atoms are able to effectively fill their outer energy levels and so achieve a more stable state. The octet rule states that atoms will react together in an attempt to fill their outer electron energy levels, and therefore reach a more stable state. For most of the lighter elements, a complete and stable outer (valence) energy level requires eight electrons (i.e. two electrons in 2s and two electrons in each of the three 2p atomic orbitals).

1.3 Summing up

1. The number of protons in an atom's nucleus is unique for each element, but the number of neutrons may vary, giving rise to isotopic forms of the same element.
2. Isotopes are very important in biology [see 'Taking it further: Isotopes in biology'].
3. Electrons are found in atomic orbitals, named 1s, 2s, 2p and so on, in increasing energy levels as they get further from the nucleus.
4. Atomic orbitals are spaces where there is a high probability of finding an electron(s); electron configurations and 'electrons in boxes' are simple ways of depicting the arrangement of electrons in an atom.
5. A maximum of two electrons occupy an s orbital or a single p orbital.
6. Elements that have full outer energy levels are stable and unreactive (see 'Taking it further: The periodic table').
7. Atoms combine (form covalent bonds) by sharing electrons or losing and gaining electrons in an attempt to fIll their atomic orbitals.

1.4 Test yourself

The answers are given on p. 179.

Question 1.1
What are the mass numbers (*A*) of the isotopes hydrogen-1, hydrogen-2 and hydrogen-3, respectively?

Question 1.2
(a) How many types of atomic orbital are present in energy level $n = 1$?
(b) How many types of atomic orbital can be present in energy level $n = 2$?

Question 1.3
Define 'atomic orbital'.

Question 1.4
(a) How many electrons are present in an atom with full 1s, 2s and 2p orbitals?
(b) Would such an atom be likely to react?

Question 1.5
(a) Arrange the following atomic orbitals in order of their energy, lowest to highest; $2p_x$, $2p_y$, $2p_z$, 2s, 1s.
(b) What does this tell you about the energy of atomic orbitals with respect to their distance from the nucleus?

Taking it further

Isotopes in biology

Isotopes of a number of elements are unstable and will readily lose mass or energy in order to reach a more stable state; these are **radioisotopes.** Radioisotopes decay, and in doing so can emit one or more forms of radiation, namely alpha particles, beta particles and gamma radiation (or gamma rays). After this decay process an entirely different element remains.

The employment of radioactive isotopes in biology probably began in 1923 at the University of Freiburg with the work of Georg Hevesy who was measuring the uptake and distribution of radioactive lead in plants (for which he later received the Nobel Prize in 1943). By the mid-1930s, the cyclotron had been invented and with it came the capability of creating artificial radioactive isotopes such as carbon-14, iodine-131, nitrogen-15, oxygen-17, phosphorus-32, sulphur-35, tritium (hydrogen-3), iron-59 and sodium-24.

There are a number of properties of radioisotopes that particularly lend them to a variety of uses in biology.

1. Radioisotopes can be detected; with the use of sophisticated monitoring equipment, we can detect, measure and follow radioisotopes in organisms and in the environment.
2. Radioisotopes of a particular element are not distinguished from the naturally occurring stable element by organisms.
3. Radioisotopes have a characteristic half-life. The half-life of a radioisotope is the time taken for the radioactivity to drop by one half; different radioisotopes decay at different rates.
4. Radioisotopes can be damaging to biological molecules.

Ecology and the environment

We can study the distribution of nutrients in ecosystems, the impact of sewage discharges on sea and land fauna, the dispersion of invertebrate pests, and many more effects, by following the distribution of incorporated radioisotopes. The radioisotopes chosen need to have a suitable half-life; long enough that the experiment can be completed, but not too long that they remain in the environment and exert possible damaging effects.

Dating of sedimentary rock

Accurate dating of sedimentary rocks can be obtained using **radiometric dating.** This uses the phenomenon of radioactive decay of isotopes. When sedimentation occurs radioactive isotopes are incorporated in the rock, and these decay to form other atoms at a known rate. This rate is measured as the **half-life** of the isotope, defined as the time taken for half the parent atoms to decay to the daughter atoms. For example, potassium-40 (^{40}K) decays to form argon-40 (^{40}Ar), which is trapped in the rocks. The amount of argon can be measured. The half-life of ^{40}K is 1.3 x 10^6 years, so it is useful for dating very old rock (as old as the Earth), the minimum age being 100 000 years.

Carbon-14 dating

Carbon-14 is continuously formed in the upper atmosphere by the action of cosmic rays on nitrogen-14. Carbon-14 is a beta emitter that decays to nitrogen-14 with a half-life of 5730 years. This isotope of carbon is often used to date the remains of anything which contains carbon. There is a naturally occurring ratio of one carbon-14 atom to about one billion carbon-12 atoms. All life is based on carbon compounds and as an organism grows it incorporates carbon-14 continuously in this ratio. When the organism dies the uptake of carbon ceases and no additional carbon-14 will be added. The concentration of the carbon-14 will decrease steadily with time by decaying to nitrogen-14. If the carbon-14 to carbon-12 ratio in a sample is measured, the age of the sample can be estimated with reasonable accuracy. The results of this technique agree to within 10% of historical records.

Isotopes in medicine

It has long been known that radiation kills cancer cells, but unfortunately it also kills other healthy cells. **Radioimmunotherapy** provides an opportunity to deliver more specific radiation to tumour cells while sparing normal tissue. The principles of radioimmunotherapy involve first making an antibody which will specifically recognise an antigen associated with the cancer cells. The antibody is then radio-labelled; this involves chemically attaching a radioisotope to the antibody. The radio-labelled antibody is then injected into the body, where it eventually 'seeks out' the cancer cells and attaches to them. Being in such close proximity to the cancer cell, the radiation from the radioisotope will kill those cells, without causing too much damage to surrounding healthy tissues.

Iodine-131 has been available for many years for treating thyroid cancer. It is now being used to attach to an antibody to provide a way of performing radioimmunotherapy. Iodine-131 has a half-life of eight days; it is a beta and gamma emitter.

The decay of iodine-131, to xenon, is shown in the equation below.

$$^{131}_{53}I \longrightarrow {}^{131}_{54}Xe + \underbrace{{}^{0}_{-1}e}_{\beta \text{ particle}} + \text{gamma radiation}$$

The high energy electron emitted in this decay, a beta particle, originates from the atomic nucleus of iodine; a neutron changes to a proton, emitting a beta particle, increasing the atomic number to 54 (= xenon), but without any change to the mass number. The iodine-131 itself is chemically linked to the antibody. The beta particles emitted by iodine-131 have sufficient energy to travel about 5 mm in biological tissues; this helps to localise the damaging effects of the beta particles to the tumour itself but also enables a significant area of cancer cells to be targeted. An additional advantage of using iodine-131 is that this radioisotope is also a gamma emitter. Gamma radiation can be detected by specialist imaging equipment. This enables doctors to learn more about the diseased tissues, particularly their extent and localisation. Medical isotope diagnostic procedures often facilitate an earlier and more complete disease diagnosis and therefore a more rapid and effective treatment.

In clinical medicine another example of an isotope-based detection method is the carbon-13 breath tests used for the detection of *Helicobacter pylori*. (Note that carbon-13 is a stable isotope, not a radioisotope.) This bacterium is responsible for causing stomach ulcers. It uses an enzyme called urease, which breaks urea down to carbon dioxide (Fig. 7). If you give the patient a 'meal' of urea that contains carbon-13, then carbon-13-labelled CO_2 ($^{13}CO_2$) will be exhaled and can be detected in the patient's breath, if the bacterium is present.

$$\underset{\text{urea}}{H_2N-\overset{*}{C}(=O)-NH_2} + H_2O \xrightarrow{\text{urease}} NH_3 + \underset{\text{carbamine acid}}{H_2N-\overset{*}{C}(=O)OH}$$

$$H_2N-\overset{*}{C}(=O)OH \longrightarrow NH_3 + {}^{*}CO_2$$

Figure 7. Conversion of urea to ammonia and carbon dioxide, catalysed by urease

The action of the enzyme urease results in the formation of ammonia and carbon dioxide; the carbon-13 (shown by the asterisk) in the urea is transferred to carbon dioxide; this can then be detected in exhaled breath.

Isotopes in research

A huge range of applications and tools are now available to the research scientist. Commonly available isotopes, such as carbon-14, iodine-131, nitrogen-15, oxygen-17, phosphorus-32, sulphur-35, tritium (hydrogen-3), iron-59 and sodium-24, can be attached to a variety of biomolecules (fats, carbohydrates, proteins, nucleic acids).

There is almost no limit to the number and kind of compounds which can be labelled and traced. We can use radio-labelled biomolecules in order to trace their metabolism and to elucidate metabolic pathways, or to deduce the mechanism of an enzyme-catalysed reaction. Isotopes used as labels have proven particularly useful in the analysis and detection of molecules. For example, consider that we might need to detect the presence of minute quantities of a specific protein. A good way to do this would be to produce and use a highly specific antibody to that protein. But how do we then specifically detect the antibody (or to be more accurate, the antibody–protein complex)? Answer, attach a radioisotope to the antibody first, then we can look for the antibody by detecting radioactivity. A relatively simple way of detecting radioactivity is through its effect on a photographic film, a technique called autoradiography.

So we can date rocks, animal and plant remains, employing naturally occurring isotopes. We can produce artificial isotopes to radio-label and follow the fate of chemicals in the environment, or in the human body. We can target radioisotopes in medicine to kill cancer cells and to localise diseased tissue. We can attach radioisotopes to biomolecules for purposes of analysis and detection.

Taking it further

The periodic table

Time line

1789
Antoine Lavoisier defines a chemical element and constructs a table of 33 of them.

1829
Johann Dobereiner announces his law of triads.

1843
Leopold Gmelin publishes his famous *Handbuch der Chemie*.

1858
Stanislao Cannizzaro assigns atomic weights to elements.

1862
Beguyer de Chancourtois makes use of atomic weights to reveal periodicity.

1865
John Newlands draws up a table of periodicity, his 'law of octaves'. Politically ridiculed, Newlands' work is not recognised until 1887.

1868
Julius Lothar Meyer constructs 'primitive' periodic table in 1864; a more sophisticated version is produced in 1868 but not published until his death in 1895.

1869
Mendeleev publishes his periodic table, based on atomic mass and chemical valency of the elements. Meyer publishes too late to claim priority over Mendeleev, but in time to confirm that the latter's discovery is based on sound chemical principles.

Historical perspectives

Since the earliest days of chemistry, attempts have been made to arrange the known elements in ways that revealed similarities between them. Ever since Antoine Lavoisier defined a chemical element and drew up a table of 33 of them for his book 'Traité élémentaire de chimie', published in 1789, there have been attempts to classify them. It took the genius of Dimitri Ivanovich Mendeleev to 'discover' the periodic table which today is so widely used and accepted. Today, the periodic table is firmly based on the properties of atomic number and the electron energy levels which surround the nucleus. Both of these concepts post-date Mendeleev by several decades; he, however, perceived them indirectly through the relative properties of atomic mass and chemical valency, and arrived at the periodic table in 1869. At this time, 65 elements were known; Mendeleev arranged these in his table, pointing out the many unoccupied positions in the overall scheme. He took the much bolder step of predicting the properties of these missing elements; moreover, the gap in atomic masses between cerium (140) and tantalum (181) suggested to him that a whole period of the table remained to be discovered. By the end of that century, most of the elements predicted by Mendeleev had been isolated, including the 'missing' period, the lanthanides.

The modern periodic table

The modern periodic table has probably reached its final form, firmly grounded on atomic theory. The periodic table shown in Fig. 8 is simple in the extent of the data it contains. Each box of the table contains the element's identity (agreed chemical symbol), the atomic mass, and the atomic number (number of protons). The elements are arranged in rows or **periods** (across), and columns or **groups** (1 to 18, down), increasing in atomic number from left to right, and in mass number. Elements in the same column have similar chemical properties. Mendeleev had noted that various properties of the elements seemed to go through cycles as the atomic number was increased. Boiling points did not simply increase with atomic number, but went through peaks and troughs; similarly for ionisation energy (the energy required to remove an electron). The number of chemical bonds that an element could form with another element varied with atomic number. Elements with atomic numbers 3, 11 or 19 could combine with only one atom of another element (Group 1 elements are said to be **monovalent**); those with atomic numbers 5 and 13 were able to form three bonds (Group 13 elements are **trivalent**). Elements with atomic numbers 2, 10, 18 etc. do not readily form bonds with any other elements; these are the stable **inert** or **noble** gases (Group 18). In other words, the properties of the elements exhibit a **periodicity**; the Periodic Law states that 'the properties of the chemical elements are a periodic function of atomic number'.

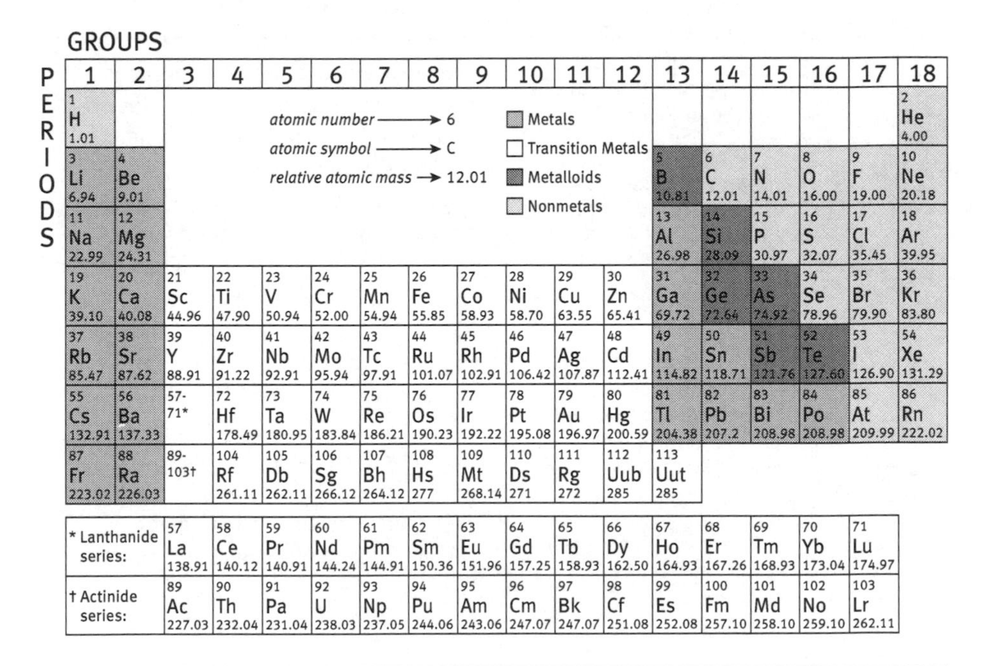

Period	1	2	3	4	5	6	7	8	9	10	11	12	13	14	15	16	17	18
1	1 H 1.01																	2 He 4.00
2	3 Li 6.94	4 Be 9.01											5 B 10.81	6 C 12.01	7 N 14.01	8 O 16.00	9 F 19.00	10 Ne 20.18
3	11 Na 22.99	12 Mg 24.31											13 Al 26.98	14 Si 28.09	15 P 30.97	16 S 32.07	17 Cl 35.45	18 Ar 39.95
4	19 K 39.10	20 Ca 40.08	21 Sc 44.96	22 Ti 47.90	23 V 50.94	24 Cr 52.00	25 Mn 54.94	26 Fe 55.85	27 Co 58.93	28 Ni 58.70	29 Cu 63.55	30 Zn 65.41	31 Ga 69.72	32 Ge 72.64	33 As 74.92	34 Se 78.96	35 Br 79.90	36 Kr 83.80
5	37 Rb 85.47	38 Sr 87.62	39 Y 88.91	40 Zr 91.22	41 Nb 92.91	42 Mo 95.94	43 Tc 97.91	44 Ru 101.07	45 Rh 102.91	46 Pd 106.42	47 Ag 107.87	48 Cd 112.41	49 In 114.82	50 Sn 118.71	51 Sb 121.76	52 Te 127.60	53 I 126.90	54 Xe 131.29
6	55 Cs 132.91	56 Ba 137.33	57-71*	72 Hf 178.49	73 Ta 180.95	74 W 183.84	75 Re 186.21	76 Os 190.23	77 Ir 192.22	78 Pt 195.08	79 Au 196.97	80 Hg 200.59	81 Tl 204.38	82 Pb 207.2	83 Bi 208.98	84 Po 208.98	85 At 209.99	86 Rn 222.02
7	87 Fr 223.02	88 Ra 226.03	89-103†	104 Rf 261.11	105 Db 262.11	106 Sg 266.12	107 Bh 264.12	108 Hs 277	109 Mt 268.14	110 Ds 271	111 Rg 272	112 Uub 285	113 Uut 285					

Series															
* Lanthanide series:	57 La 138.91	58 Ce 140.12	59 Pr 140.91	60 Nd 144.24	61 Pm 144.91	62 Sm 150.36	63 Eu 151.96	64 Gd 157.25	65 Tb 158.93	66 Dy 162.50	67 Ho 164.93	68 Er 167.26	69 Tm 168.93	70 Yb 173.04	71 Lu 174.97
† Actinide series:	89 Ac 227.03	90 Th 232.04	91 Pa 231.04	92 U 238.03	93 Np 237.05	94 Pu 244.06	95 Am 243.06	96 Cm 247.07	97 Bk 247.07	98 Cf 251.08	99 Es 252.08	100 Fm 257.10	101 Md 258.10	102 No 259.10	103 Lr 262.11

Figure 8. The modern periodic table

Valence and the octet rule

Atoms whose outer (**valence**) electron energy levels hold a complete set of electrons are substantially more stable than those which do not. The electronic configuration of sodium (Na) is $1s^2 2s^2 2p^6 3s^1$. If sodium were to lose one electron, and form the Na^+ ion, it would gain the electronic configuration of the stable element neon ($1s^2 2s^2 2p^6$). That such an ionisation can occur is supported by the relatively low ionisation energy for sodium (and other Group 1 elements). Similarly, chlorine has the electronic configuration $1s^2 2s^2 2p^6 3s^2 3p^5$; in gaining one electron chlorine would attain the electronic structure of the stable element argon ($1s^2 2s^2 2p^6 3s^2 3p^6$). Hence, the Cl^- ion is readily formed. For most of the lighter elements a complete outer or valence shell requires eight electrons.

Consequently, the observations noted above are said to obey the **octet rule**.

$$Na \longrightarrow Na^+ + e^-$$

$$1s^2 2s^2 2p^6 3s^1 \qquad 1s^2 2s^2 2p^6$$

$$Cl + e^- \longrightarrow Cl^-$$

$$1s^2 2s^2 2p^6 3s^2 3p^5 \qquad 1s^2 2s^2 2p^6 3s^2 3p^6$$

The properties of the elements exhibit **trends**. These trends can be predicted using the periodic table and can be explained and understood by considering the electron configurations of the elements. Elements tend to gain or lose valence electrons to achieve a stable octet formation. Stable octets are seen in the inert gases, Group 18 of the periodic table. In each group the elements have the same outer energy level electron configuration, and therefore the same **valency**. For example, Group 1 elements require the input of a relatively small amount of energy to lose an electron, i.e. to ionise. In contrast it is extremely difficult to remove an electron from a Group 17 element; these are close to a stable octet and complete outer electron energy level and therefore have a strong tendency to acquire electrons. Electrons are added one at a time, moving from left to right across a period. As this happens, across a row the electrons in the outermost energy level experience increasingly strong nuclear attraction and become closer and more tightly bound. Furthermore, moving down a group, the outermost electrons become less tightly bound to the nucleus. This happens because the number of filled principal energy levels increases downward in each group, and so the outermost electrons are more shielded from the attraction of the nucleus. These trends explain the periodicity observed with respect to atomic radius, ionisation energy, electron affinity and electronegativity in moving from left to right across a period.

Atomic radius

The atomic radius (defined as half the distance between the centres of two neighbouring atoms) decreases across a period from left to right and increases down a given group. From left to right across a period, electrons are added one at a time to the outer energy shell. Electrons within the same energy level cannot shield each other from the nuclear attraction. Since the number of protons is also increasing, the effective nuclear charge increases across a period. Moving down a group, the number of electrons and filled electron energy levels increases, but the number of valence electrons stays the same. The outermost electrons are exposed to approximately the same effective nuclear charge, but are found further from the nucleus as the number of filled energy levels increases; therefore, the atomic radii increase.

Ionisation energy

The ionisation energy, or ionisation potential, is the energy required to completely remove an electron from a gaseous atom or ion. The closer and more tightly bound an electron is to the nucleus, the more difficult it will be to remove. Ionisation energies increase moving from left to right across a period (decreasing atomic radius), and decrease moving down a group (increasing atomic radius). Group 1 elements have low ionisation energies, with the loss of an electron forming a stable octet.

Electron affinity

Electron affinity reflects the ability of an atom to accept an electron. Atoms with stronger effective nuclear charge have greater electron affinity. For example, Group 17 elements, the halogens, have high electron affinities; the addition of an electron to such elements results in a completely filled outer electron energy level.

Electronegativity

Electronegativity is a measure of the attraction of an atom for the electrons in a chemical bond. The higher the electronegativity of an atom, the greater its attraction for bonding electrons. Electronegativity is related to ionisation energy. Atoms with low ionisation energies have low electronegativities, because their nuclei do not exert a strong attractive force on the electrons, and *vice versa*. In a period, moving from left to right, electronegativity increases (as ionisation energy increases). Going down a group in the periodic table, electronegativity decreases as both atomic number and atomic radius increase, and ionisation energy decreases.

Biological life

About 25 of the 92 natural elements are known to be essential to life. Carbon, oxygen, hydrogen and nitrogen make up 96% of living matter. Phosphorus, sulphur, calcium, potassium and a few other elements account for most of the remaining 4% of an organism's weight. Biomolecules are constructed using carbon to provide the framework, and hydrogen, nitrogen and oxygen to provide functionality. Sulphur and phosphorus are important constituents of some biomolecules. Trace elements are required by organisms in only minute quantities. Trace metals, such as iron, magnesium and zinc, are important catalytic components present in enzymes.

02 Bonding, electrons and molecules

BASIC CONCEPTS:
Understanding the nature of covalent bonding is an essential prerequisite to predicting the behaviour of biological molecules. Here we explore the nature of covalent bonding and types of covalent bonds.

Atoms react together to form molecules. Biological life forms are able to produce complex and extremely large molecules. Inherent in the stability of such molecules are the **intramolecular forces** between atoms which hold the molecules together, producing bonds which are strong and relatively stable. Such bonds may be referred to as covalent, dative covalent, polar covalent or ionic.

2.1 What is a covalent bond?

A covalent bond is formed between two atoms, commonly by the sharing of two electrons, one electron being donated by each of the two atoms. Consider the hydrogen atom. Remember hydrogen has one proton in the nucleus and one electron in a 1s orbital.

Using our diagrams of atomic orbitals, we can consider the hydrogen molecule, H_2, as being formed from the overlap of two 1s orbitals, one from each hydrogen atom (Fig. 9). In this way, each hydrogen atom effectively has two electrons at any time. In other words, each hydrogen atom has a full 1s orbital. Furthermore, since both hydrogen atoms in this bond are identical, we can assume that the electrons are shared equally between the two atoms, producing a symmetrical covalent bond. This 'head-to-head' merging of atomic orbitals forms **sigma molecular orbitals**. It is the formation of bonding molecular orbitals which results in covalent bonding.

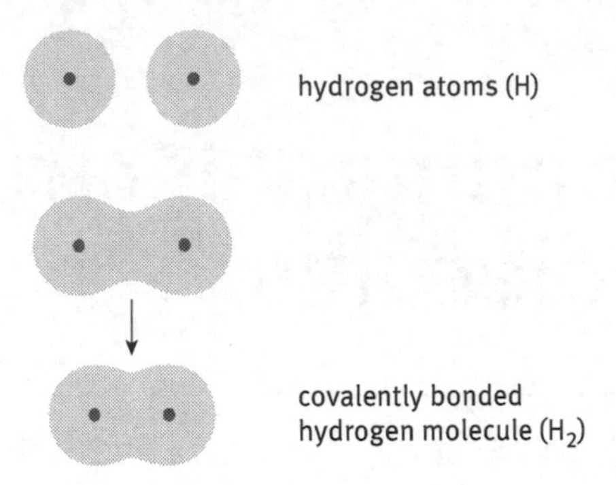

Figure 9. Formation of a covalent bond in the hydrogen molecule

QUESTION

How can two electrons fill the 1s orbital of both H atoms? Remember, orbitals represent spaces where there is a high probability of finding an electron, and electrons move close to the speed of light. So, statistically there is a high probability that at any one time both electrons will be associated with one or other hydrogen atom!

Sigma molecular orbitals are just as easily formed between p orbitals, again in a 'head-to-head' merging (Fig. 10).

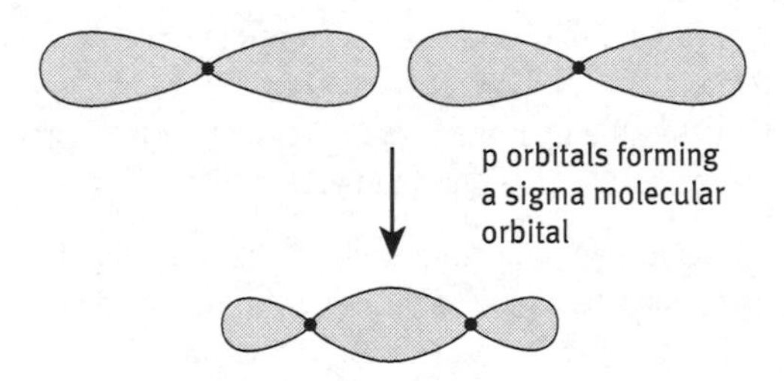

Figure 10. Sigma molecular orbital formation between p orbitals

QUESTION

Why do atoms form covalent bonds?

In the hydrogen molecule, both H atoms effectively gain a full 1s orbital, and therefore attain a stable state. For hydrogen this simply means filling its 1s orbital. In the case of nitrogen, this means filling its three 2p orbitals (see Fig. 4). Atoms whose outer (valence) electron energy level holds a complete set of electrons are substantially more stable than those which do not. By sharing electrons through covalent bonding, atoms are able to effectively fill their outer electron energy levels and so gain greater stability. For most of the lighter elements a complete outer or valence shell requires eight electrons (two electrons in the 2s orbital plus six electrons in the three 2p orbitals).

TAKING IT FURTHER: The periodic table (p. 13)

2.2 Non-bonding electrons – lone pairs

In bonding, atoms attempt to attain a complete set of electrons in their valence shells as this arrangement generally leads to stability. In many molecules this means that atoms have electrons in their valence shells which are not involved in bonding to other atoms. These non-bonded electrons are known as **lone pairs**.

REMINDER

Atoms share electrons in covalent bonds in order that each attains a full outer electron energy level, and hence greater stability

Consider what happens when two fluorine atoms react together to form a molecule (Fig. 11). The electron configuration of fluorine is $1s^2 2s^2 2p^5$ and so each atom of fluorine has seven outer electrons. When the half-filled p orbitals of the fluorine atoms, each containing a single electron, overlap, a sigma bond is formed. By sharing electrons in this way each fluorine atom now has a filled outer valence shell of eight electrons. However, six of the electrons on each fluorine atom are not involved in bonding. They are paired together in non-bonding orbitals and are called lone pairs. Lone pairs of electrons have an effect upon the reactivity of a molecule and also upon its shape.

In the water molecule the two lone pairs and two bonding pairs of electrons on the oxygen atom are arranged tetrahedrally. This is the arrangement in which the negative centres experience the minimum repulsion. The lone pairs exert a larger repulsive effect than the bonding pairs and so the hydrogen atoms are pushed together slightly and the bond angle becomes slightly less than the tetrahedral angle, 109°. Overall the molecule is therefore V-shaped (Fig. 11).

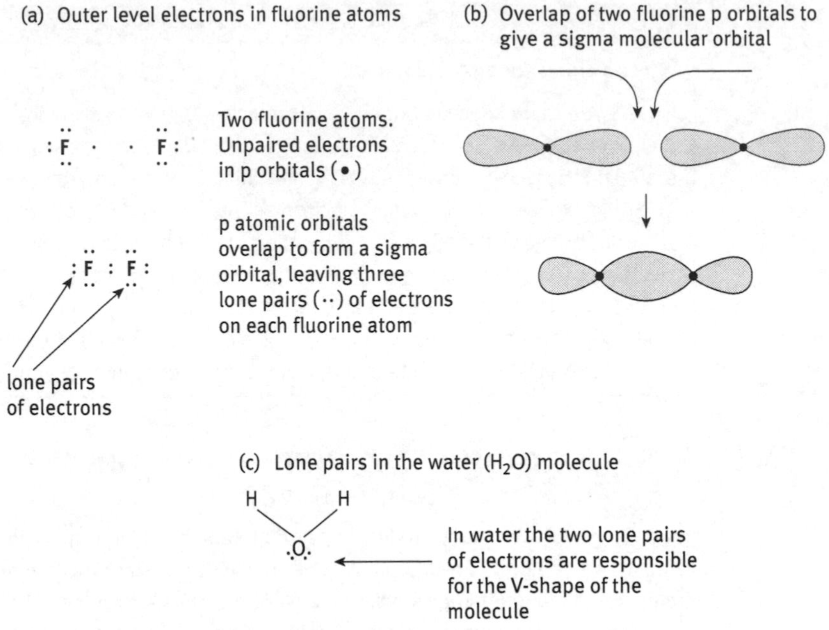

(a) Outer level electrons in fluorine atoms

:F· ·F:

Two fluorine atoms. Unpaired electrons in p orbitals (•)

:F:F:

p atomic orbitals overlap to form a sigma orbital, leaving three lone pairs (··) of electrons on each fluorine atom

lone pairs of electrons

(b) Overlap of two fluorine p orbitals to give a sigma molecular orbital

(c) Lone pairs in the water (H_2O) molecule

H H
O

In water the two lone pairs of electrons are responsible for the V-shape of the molecule

Figure 11. Lone pairs of electrons

2.3 Pi molecular orbitals

In addition to forming sigma molecular orbitals, electrons in p orbitals can also overlap in a 'side-to-side' fashion, forming pi molecular orbitals (Fig. 12).

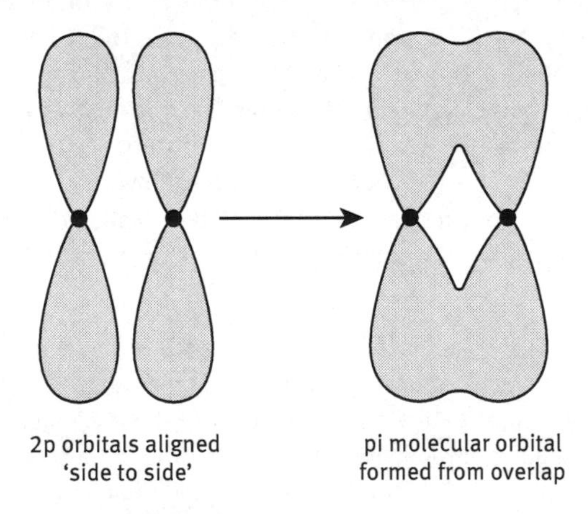

Figure 12. Formation of pi molecular orbitals

Pi bonding between atoms occurs in addition to, rather than instead of, sigma bonding. When both a sigma molecular orbital and a pi molecular orbital are formed between two atoms, a 'double bond' results. Whereas sigma bonding, shown by shorthand as a single line connecting two atoms, allows complete rotation about the bond, pi bonding, shown in shorthand as a double line between atoms (one sigma bond plus one pi bond) restricts rotation about that bond.

Yes – free rotation about bond

No – rotation about bond is restricted

single bond

double bond

Pi bonding has important consequences in determining the shapes (conformations) of biological molecules, and particularly protein molecules.

TAKING IT FURTHER: The peptide bond (p. 28)

REMINDER

Covalently bonded atoms can rotate freely about a sigma molecular orbital, but not about a pi molecular orbital

2.4 Coordinate bonds

A covalent bond is formed by two atoms sharing a pair of electrons. The atoms are held together because the electron pair is attracted by both of the nuclei. In the formation of a simple covalent bond, each atom supplies one electron to the bond, but that doesn't have to be the case. A **coordinate** bond (also called a **dative** covalent bond) is a covalent bond (a shared pair of electrons) in which both electrons come from the same atom. In simple diagrams, a coordinate bond is shown by an arrow. The arrow points from the atom donating the lone pair to the atom accepting it. In Fig. 13, the nitrogen atom in the ammonia molecule is donating its pair of electrons to the empty 1s orbital of the positive hydrogen ion, a proton. A new dative covalent bond is formed in the ammonium ion, NH_4^+. Once formed, each of the N-H bonds is equivalent, irrespective of the source of the electrons.

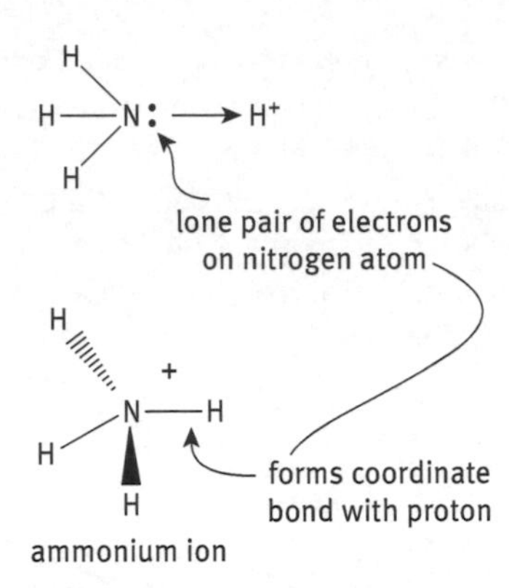

Figure 13. Coordinate covalent bond formation in the ammonium ion

2.5 Electronegativity and polar covalent bonds

TAKING IT FURTHER:
The periodic table
(p. 13)

In covalent bonds between like atoms, electrons are shared equally between the two atoms in the bond. However, certain types of atoms are able to exert a greater pull on electrons than others. This ability to attract electrons within a bond is called the **electronegativity** of that element. The electronegativity of an atom is a property dependent on the size of the atom, and the degree of 'shielding' of the positively charged nucleus by the negatively charged electrons. The section of the periodic table below (Fig. 14) shows the relative electronegativities of a number of atoms; the larger the value the greater their electronegativity.

H 2.2						
Li 1.0	Be 1.5	B 2.0	C 2.5	N 3.1	O 3.5	F 4.1
Na 0.9	Mg 1.2	Al 1.5	Si 1.7	P 2.1	S 2.4	Cl 2.8
K 0.8	Ca 1.0	Ga 1.8	Ge 2.0	As 2.2	Se 2.5	Br 2.7

Figure 14. Electronegativity values of some elements of the periodic table

In general, elements on the right-hand side of the periodic table are more electronegative than those on the left-hand side. In addition, as atoms get bigger (i.e. going down a group in the periodic table), electronegativity decreases [see 'Taking it further: The periodic table']. In biological molecules, oxygen and nitrogen are particularly important; their electronegativity has important consequences in terms of the reactivity and associations of molecules.

REMINDER

Electronegativity is a measure of the degree to which an atom 'draws' electrons towards itself in a bond

2.6 What effect does electronegativity have on covalent bonds?

In a sigma molecular orbital between carbon and oxygen, both atoms contribute one electron to the bond. Oxygen draws electrons towards itself because it is relatively more electronegative than carbon. Consequently, the two electrons in the bond are more likely to be found closer to the oxygen atom than the carbon atom (Fig. 15). This unequal electron distribution results in the formation of a **polar covalent bond**.

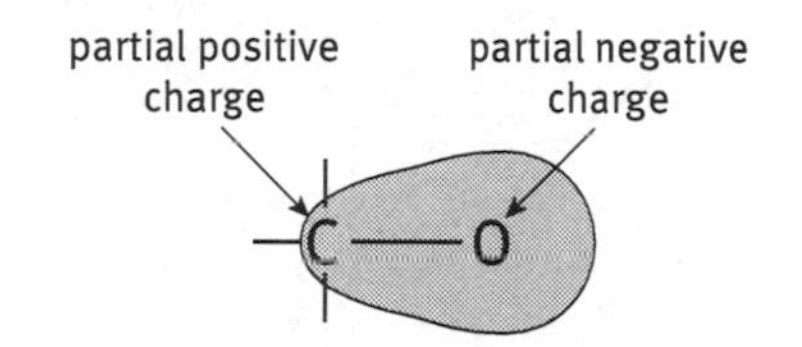

Figure 15. Unequal electron distribution in a polar covalent bond

REMINDER

A **polar covalent bond** is a distorted covalent bond in which there is an unequal distribution of electrons

This unequal electron distribution results in one end of the bond being slightly positive (electron deficient) and the other end being slightly negative, and thus a **dipole** is produced.

REMINDER

A **dipole** is produced when a pair of electric charges, of equal magnitude but opposite polarity, are separated by some (usually small) distance

Polar covalent bonds play an important role in biological molecules. They invariably form the basis of **functional (reactive) groups** on biological molecules. Such groups are responsible for the reactivity between molecules, their solubility in water and contribute to the **intermolecular forces** between molecules.

2.7 Ionic bonds

Ionic bonds occur when there is a complete transfer of electron(s) from one atom to another resulting in two ions, one positively charged and the other negatively charged. Electrons are not 'shared' as in a covalent bond, but are 'lost' or 'gained'. For example, when a sodium atom (Na) donates its one electron in its outer 3s electron energy level to a chlorine (Cl) atom, which needs one electron to fill its outer 3p electron shell, sodium chloride results. The bond between the two new ions (Na^+Cl^-) is an ionic bond.

When sodium loses an electron from the 3s atomic orbital, all available atomic orbitals in the $n = 2$ energy level are filled, containing a total of eight electrons ($2s^2 + 2p^6 = 8$), and the sodium ion, Na^+, has reached a more stable state.

$$\begin{array}{ccc} \text{Na} & \rightarrow & \text{Na}^+ + e^- \\ 1s^2 2s^2 2p^6 3s^1 & & 1s^2 2s^2 2p^6 \end{array}$$

When chlorine gains an electron, all available atomic orbitals in the $n = 3$ energy level are filled, containing a total of eight electrons ($3s^2 + 3p^6 = 8$), and the chloride ion, Cl^-, has reached a more stable state.

$$\begin{array}{ccc} \text{Cl} + e^- & \rightarrow & \text{Cl}^- \\ 1s^2 2s^2 2p^6 3s^2 3p^5 & & 1s^2 2s^2 2p^6 3s^2 3p^6 \end{array}$$

As with covalent bond formation, the 'driving force' in forming ionic bonds is to achieve a full, and therefore stable, outer electron energy level. In ionic bond formation it is the ion, rather than the atom, that reaches a more stable state and electrons are lost or gained, rather than shared, to achieve this state.

2.8 The concept of the chemical bond

We have described the covalent bond, the coordinate bond, the polar covalent bond and the ionic bond. In reality, the intramolecular forces between atoms lie within a spectrum of bonding, with covalent and ionic representing extreme ends of this spectrum. The type of bonding which prevails depends upon the electronegativity difference between the atoms involved. Pure covalent bonds exist where the electronegativity difference lies between 0 and 0.4, polar covalent bonds where the electronegativity difference is between 0.4 and 2.1, and ionic bonds where the electronegativity difference is greater than 2.1 (Fig. 16).

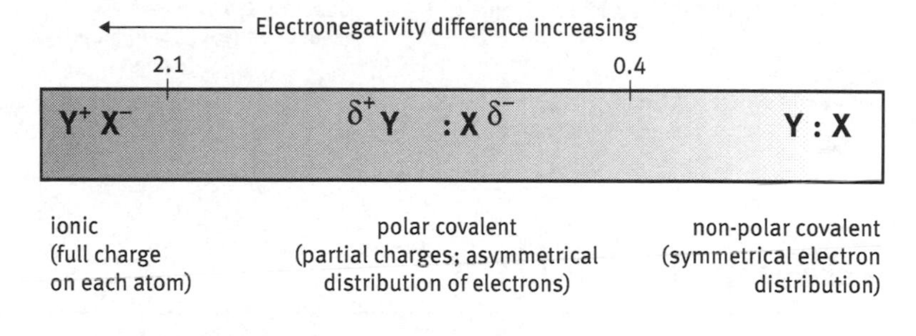

Figure 16. A spectrum of bonding between atoms

The intramolecular forces between atoms provide the inherent stability for molecular construction. However, central to biological life are the rapid and transient interactions that must occur between different molecules. Such intermolecular forces provide the basis for interaction and recognition at the molecular level of life, and are dealt with in Chapter 3.

2.9 Summing up

1. A covalent bond is formed by the sharing of two electrons between two atoms, normally one electron being provided by each atom.
2. Coordinate (dative covalent) bonds are formed where both electrons in the bond are provided by just one of the atoms.
3. Atoms form covalent bonds in an attempt to fill their outer atomic orbitals and so reach a more stable state.
4. Both sigma ('head-to-head'), and pi ('side-to-side') molecular orbitals are possible through the merging of atomic orbitals.
5. Rotation about a double covalent bond is restricted (see 'Taking it further: The peptide bond').

6. Some atoms are more electronegative than others, which can result in the formation of polar covalent bonds.
7. Polar covalent bonds form the basis of functional groups on molecules.
8. Ionic bonding involves the gain or loss of electrons, the ions so formed achieving stability through attaining full outer atomic orbitals.

2.10 Test yourself

The answers are given on p. 179.

Question 2.1
Describe the relationship between atomic orbitals, molecular orbitals and covalent bonds.

Question 2.2
Define the difference between a sigma molecular orbital and a pi molecular orbital.

Question 2.3
What do you understand by a 'polar covalent bond'?

Question 2.4
What is unusual about a dative covalent bond?

Question 2.5
What do you understand by the 'octet rule'?

Taking it further

The peptide bond

Proteins are polymers of amino acids. Amino acids are linked together by the formation of peptide bonds, between the carboxyl group of one amino acid and the amine group of another amino acid, in a condensation reaction (Fig. 17). A **condensation reaction** is a chemical reaction in which a molecule of water is lost.

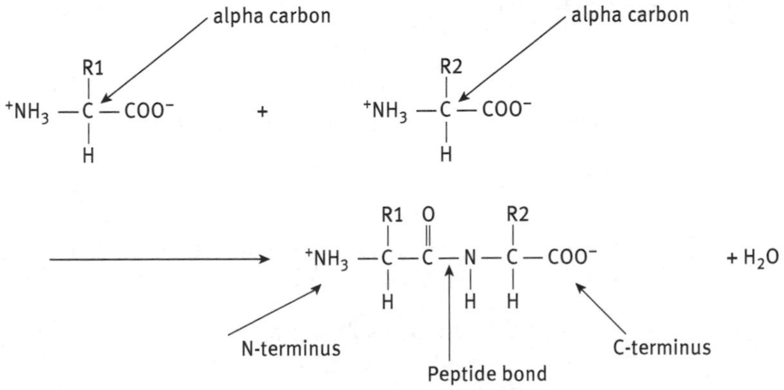

$$^{+}NH_3-C(R1)(H)-COO^- + {}^{+}NH_3-C(R2)(H)-COO^- \longrightarrow {}^{+}NH_3-C(R1)(H)-C(=O)-N(H)-C(R2)(H)-COO^- + H_2O$$

Figure 17. Formation of a peptide bond between two amino acids in a condensation reaction

The sequence of amino acids in a protein determines the primary structure of that protein. The flexibility and folding of the polypeptide chain is responsible for the specific three-dimensional shape of the protein, which is the main determinant of the structure–activity characteristics of that protein. Proteins derive their name from the ancient Greek sea-god **Proteus** who could change shape; the name acknowledges the many different properties and functions of proteins.

We have seen that there is free rotation about a single sigma covalent bond, but that rotation is restricted about a double (sigma + pi) covalent bond. The peptide bond is special in the sense that it is a partial double bond, and so rotation about the bond is restricted. This property of the peptide bond has a profound effect in determining the conformation of the polypeptide chain.

The partial double bond character of the peptide bond

The partial double bond character of the peptide bond (Fig. 18) is a consequence of the electronic configuration of the nitrogen atom, and of the pi bonding in the carbonyl group.

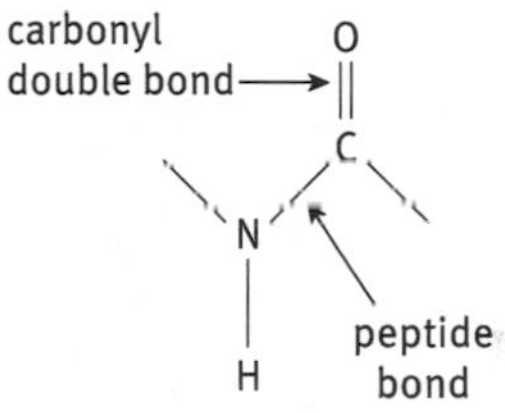

Figure 18. The peptide bond

Nitrogen (N), atomic number 7, has the electronic configuration

$1s^2\ 2s^2\ 2p^3$

Prior to covalent bonding, the atomic orbitals in the $n = 2$ energy level of nitrogen are hybridised. (Hybridisation is covered further in Section 5.2.)

Hybridisation of atomic orbitals in the same energy level occurs in order to maximise the number of covalent bonds which can be formed. The more covalent bonds which can be formed, the more stable is the atom. Nitrogen is capable of a number of types of hybridisation; in the case of peptide formation the hybridisation is referred to as sp^2 hybridisation; the s refers to the 2s orbital, and p^2 to the fact that two of the three p orbitals are hybridised. We can see this using 'electrons in boxes' diagrams (Fig. 19).

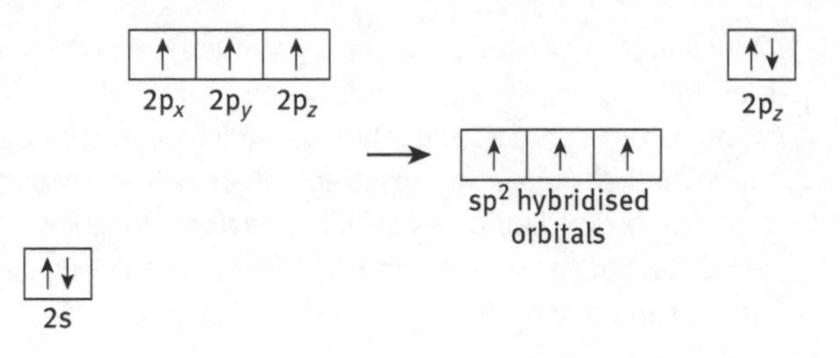

Figure 19. Hybridisation of atomic orbitals in a nitrogen atom

Electrons in the 2s orbital are raised to the slightly higher energy level of the 2p atomic orbitals; three of the resulting four orbitals are hybridised (sp^2), leaving one p orbital unchanged.

Drawing the resulting atomic orbitals (Fig. 20) we can see that the three sp^2 hybridised orbitals are available to form sigma covalent bonds, leaving the p orbital above and below the plane of the page (the three sp^2 hybrid orbitals are arranged in a planar orientation).

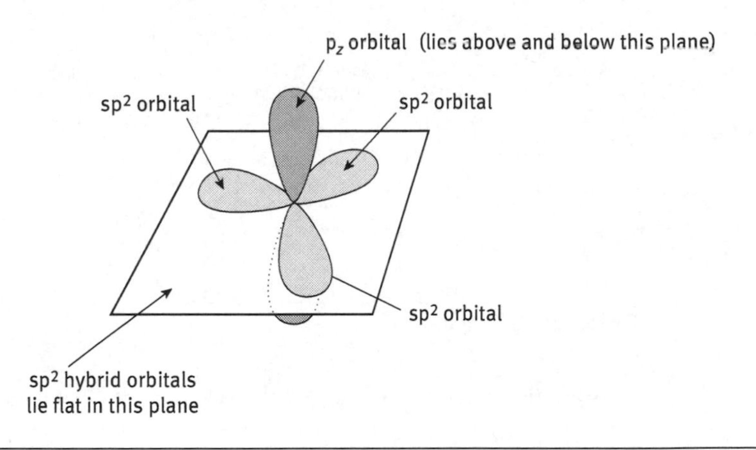

Figure 20. Spatial arrangement of sp^2 hybrid orbitals

In Fig. 21 we have shown just the p orbitals in the peptide bond. The p orbital on the nitrogen atom is close enough to interact with the p orbitals on the carbon and oxygen atoms of the carbonyl group. The 'side-to-side' merging of p orbitals forms pi covalent bonds.

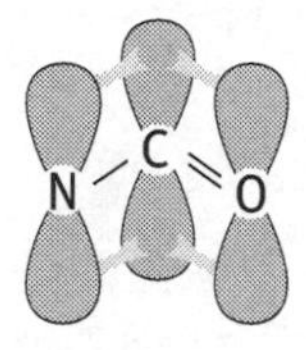

Figure 21. p orbitals of the peptide bond

Electrons can move between the p orbitals (electrons are said to be delocalised) in the formation of pi covalent bonds. In fact, the peptide bond is a resonance structure, shown in Fig. 22.

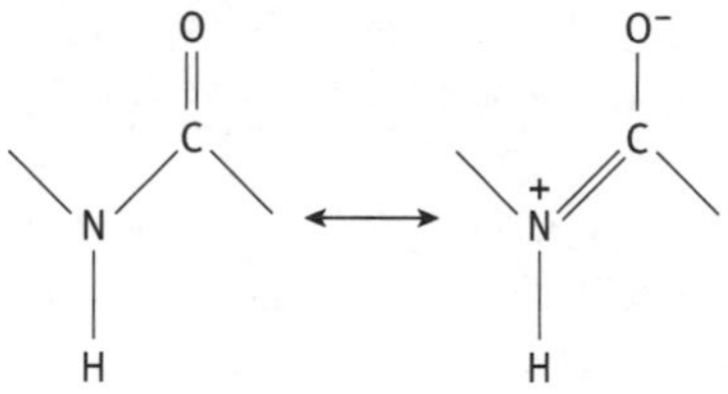

Figure 22. The peptide bond is a resonance structure

So we see that the peptide bond has a partial double bond character; there is a variable amount (about 40%) of pi bonding between the nitrogen and carbon, sufficient to restrict bond rotation.

The important properties and outcomes of the peptide bond can be summarised as follows.

- The rigidity of the peptide bond reduces the degrees of freedom of the polypeptide during folding.
- Due to the double bond character, the six atoms involved in the peptide bond group are always planar (Fig. 23).

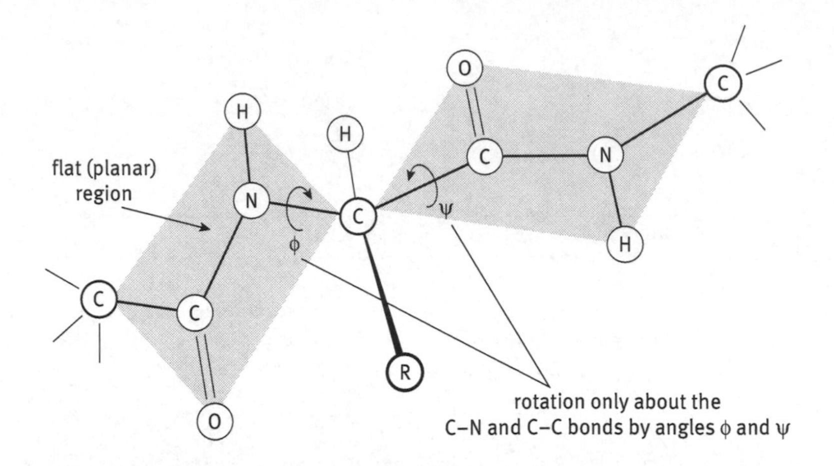

Figure 23. Atoms around a peptide bond lie in a planar conformation

- Therefore, rotation about the C–N and C–C bonds by angles of φ and ψ respectively defines the shape of the polypeptide. If all psi and phi angles are the same the peptide assumes a repeating structure. For certain combinations of angles this can take the form of a helical structure (the alpha-helix) or a beta-sheet structure (Fig. 24).

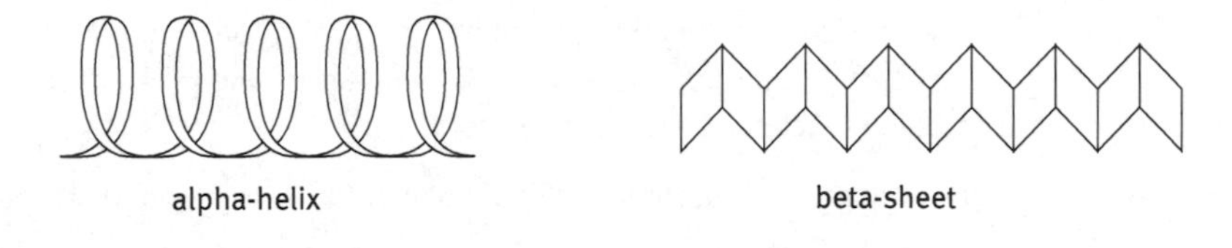

Figure 24. Alpha-helices and beta-sheets are common protein conformations

Furthermore:

- The peptide bond is invariably found in the ***trans*** conformation, i.e. alpha-carbon atoms are on **opposite** sides of the C-N peptide bond (Fig. 25). This avoids steric hindrance between groups attached to the alpha-carbon atoms (this is a form of isomerism; see Section 6.3).

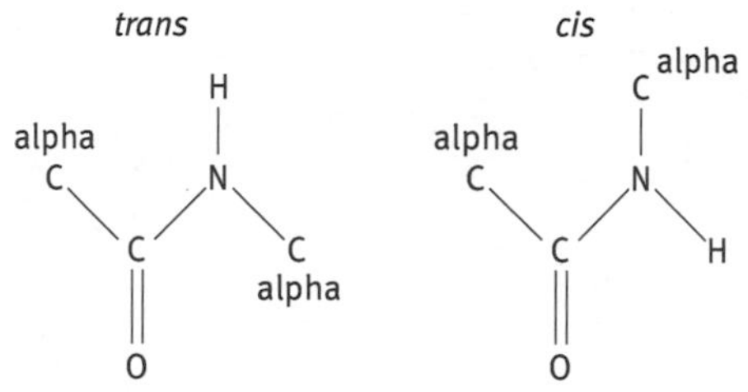

Figure 25. *cis* and *trans* conformations of the peptide bond

- The resonance donation of electrons by the nitrogen atom makes the carbonyl less electrophilic (electron-seeking). As a result the amide is comparatively unreactive. This is a good thing otherwise proteins would be too reactive to be of much use in biological systems.

The 'special' amino acid proline, a secondary amine, as opposed to other naturally occurring amino acids that are all primary amines, is unusual in that its amino group forms part of a rigid and planar ring structure (Fig. 26).

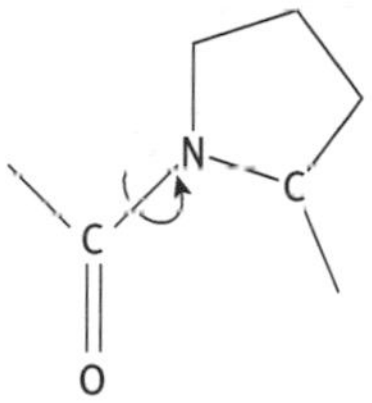

Figure 26. Rotation is possible about a proline peptide bond

Proline forms peptide bonds with other amino acids, but because of the ring structure, the peptide bond so formed lacks a partial double bond character. Thus, rotation about this peptide bond is possible. Therefore, when proline is incorporated into a polypeptide chain (Fig. 27), rotation about its peptide bond, plus lack of rotation about the phi -N-C- bond (now part of the ring structure of proline), causes a 'kink' in the polypeptide chain. The presence of proline inhibits alpha- and beta-chain conformations in proteins; proline is also found at the 'bends' in polypeptide chains.

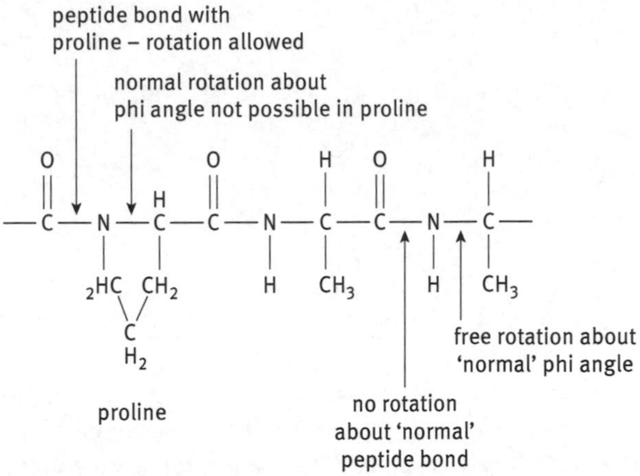

peptide bond with
proline – rotation allowed

normal rotation about
phi angle not possible in proline

proline

free rotation about
'normal' phi angle

no rotation
about 'normal'
peptide bond

Figure 27. Peptide bonds in a polypeptide chain

Thus we can see, at the level of the primary structure of proteins, how the special characteristics of the peptide bond are so important in constraining the three-dimensional conformations of polypeptide chains.

03 Interactions between molecules

BASIC CONCEPTS:

Covalent bonds can be thought of as the 'glue' which holds molecules together; they are strong, stable and require significant energy to break. On the other hand, intermolecular interactions are relatively weak, and in water form and break rapidly in a reversible manner. Weak though they are, intermolecular interactions are collectively strong and vitally important in understanding the interactions of biological molecules.

Every biological event involves molecular interactions. Molecules must interact (bind) to initiate an action, and then separate. Whether it is an enzyme binding to its substrate, a hormone binding to its receptor on a membrane, or RNA being transcribed on a DNA template, all require to be bound in a precise orientation for a short time. Such 'recognition' events are made possible through intermolecular interactions.

REMINDER

Intermolecular interactions are weak, but collectively strong

Intermolecular interactions may be broadly classified as **electrostatic** or **hydrophobic** in nature. **Electrostatic interactions** include a number of types that we can classify as hydrogen bonding, charge–charge interactions, and short-range electrostatic interactions (van der Waals forces).

3.1 Hydrogen bonding

This is an important, and relatively strong, form of intermolecular interaction. A 'classic' example of hydrogen bonding occurs between water molecules (Fig. 28).

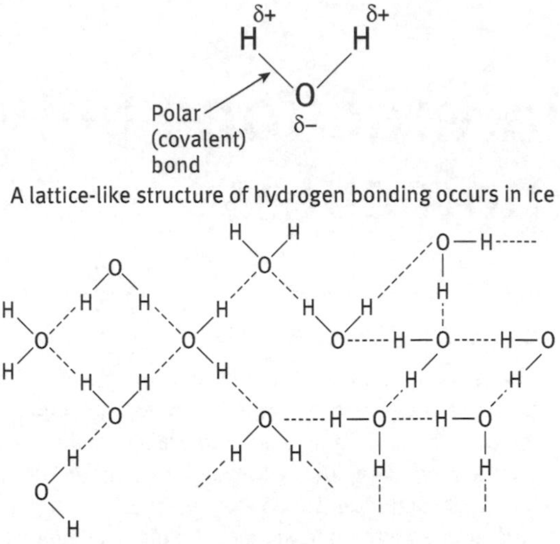

Polar (covalent) bond

A lattice-like structure of hydrogen bonding occurs in ice

Figure 28. Hydrogen bonding in water and ice

In water, the electronegative oxygen draws electrons towards itself, causing the two hydrogen atoms to carry a slight positive charge. There is a large dipole within the O – H bonds. The oxygen atom of the water molecule possesses two **lone pairs** of electrons whereas the hydrogen atoms have lost much of their share of the electrons to the electronegative oxygen atom. Hydrogen bonding involves the oxygen atom of one molecule of water sharing a lone pair of electrons with the slightly positively charged hydrogen atom of another water molecule. Water molecules therefore interact with each other through hydrogen bonding. In liquid water, such interactions rapidly form and break, but in ice a lattice-like arrangement is formed. This property of water is paramount in determining its unique characteristics as the 'solvent of life'.

Hydrogen bonds can form whenever a strongly electronegative atom (e.g. oxygen, nitrogen and fluorine) approaches a hydrogen atom which is covalently attached to a second strongly electronegative atom.

For example, hydrogen bonding can occur between a carbonyl and amino group.

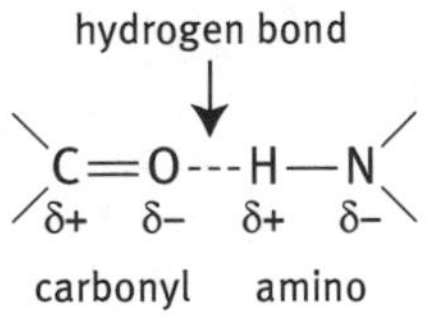

hydrogen bond

$C=O \cdots H-N$

$\delta+ \quad \delta- \quad \delta+ \quad \delta-$

carbonyl amino

A variety of such groups, containing an electronegative atom, occur in biological molecules.

hydroxyl
in alcohols, carbohydrates, proteins and nucleotides

$—O^{\delta-}—H^{\delta+}$

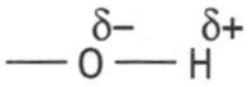

carbonyl
in organic acids, carbohydrates, proteins and nucleotides

$O^{\delta-}$ double-bonded to $C^{\delta+}$

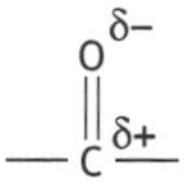

amino
in proteins and nucleotides

$H^{\delta+}$ bonded to $N^{\delta-}$

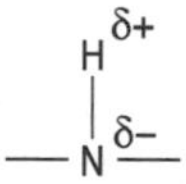

In every hydrogen bond there is the acceptor, a hydrogen atom attached to an electronegative atom, and the donor, an electronegative atom (O, N or F) with one or more lone pairs of electrons.

REMINDER

Hydrogen bonds can form whenever a strongly electronegative atom (O, N or F) approaches a hydrogen atom that is covalently attached to a second strongly electronegative atom

The hydrogen bond is of particular importance in biological systems. It may look like a simple charge–charge interaction, but the hydrogen bond has special characteristics. With this type of intermolecular interaction the optimum hydrogen bond has a distance of about 2.8 Å and is linear (an ångström, Å, is 1×10^{-10} metre, of the order of the size of an atom). For example, in the hydrogen bond between a carbonyl and amino group shown above, we could draw a straight line between the oxygen and the nitrogen, along which the bond would be aligned. In other words, the direction of the bond is relatively constrained. Compared to other intermolecular interactions, the hydrogen bond is also relatively strong. These properties of the hydrogen bond have important consequences in biology, constraining as they do the distances and orientations between molecules. Hydrogen bonds are significant in determining the structures of proteins, and in holding together the double-stranded DNA molecule.

REMINDER

Hydrogen bonds have both an optimum length and directionality

Hydrogen bonds occur between the purine and pyrimidine bases of opposite DNA strands, constraining the bases to lie flat (planar) and at a defined distance from one another. Furthermore, and because of hydrogen bonding, only thymine will bond with adenine, and only cytosine with guanine (Fig. 29).

So, our genetic code is determined by hydrogen bonding!

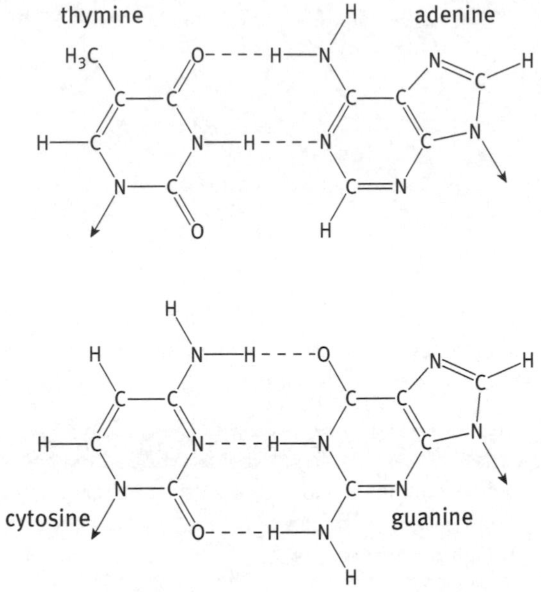

Figure 29. Hydrogen bonding between DNA bases

3.2 Charge–charge interactions

Charge–charge interactions are electrostatic in nature (as indeed are hydrogen bonds) but, unlike hydrogen bonds, the distance between interacting groups is not 'fixed' and the alignment of the interaction is not so important. Any group or atom in a molecule which carries a charge, as a result of either the electronegativity difference between atoms, the presence of an acidic or basic group, or an atom with a lone pair of electrons, can engage in charge–charge interactions. That interaction may be attractive (with an oppositely charged group) or repulsive (with a similarly charged group). Both are important in the interaction of biological molecules.

Groups which commonly carry a charge at physiological pH, and which occur particularly in proteins, are the carbonyl (negatively charged) and the amino (positively charged) groups. The charge on groups which have acidic and basic characters are necessarily very dependent on the pH of the solvent. Low pH (acid, with a higher H^+ concentration) leads to protonation; for example the amino group NH_2 exists as NH_3^+. A higher pH (more alkaline) favours dissociation of groups, for example the carboxyl group COOH exists as COO^-.

Proteins consist of amino acids, linked together through peptide bonds to produce long polypeptide chains (Fig. 30). The R groups (side groups, e.g. R1 and R2 in Fig. 30) of amino acids in the chain frequently contain carboxyl

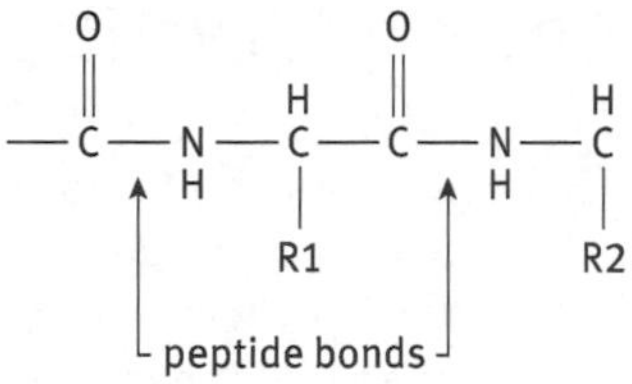

Figure 30. Side groups in polypeptide chain

(-COOH) or amino (-NH_2) groups. At physiological pH, the carboxyl group dissociates, losing a proton (H^+) and becoming negatively charged, -COO^-. At physiological pH, the amino group associates with a proton to become positively charged (-NH_3^+). On cellular proteins, therefore, there are likely to be numerous charged groups, where charge–charge interactions become possible between molecules, or indeed within molecules.

REMINDER

Interaction between charged groups is likely to be very pH-dependent

3.3 Short range charge–charge interactions

At very short distances between molecules, another type of charge–charge interaction becomes possible, referred to as **van der Waals forces**. These forces are very weak attractions (or repulsions) that occur between atoms or molecules at close range. In a covalent bond between atoms, the two electrons in that bond will, for most of the time, be found equally distributed between the two atoms. But electrons are not fixed. Statistically, there will be times when both electrons are next to one or other of the atoms; this will momentarily make that atom negatively charged, leaving the other atom in the bond positively charged, and so we will have a **temporary dipole**. (The distribution of charges in a water molecule is an example of a **permanent dipole**.) Alternatively, if we approach a covalent bond with, say, a positively charged amino group, then we could **induce** a dipole in that bond, by drawing electrons towards one end of the bond. This constant movement and redistribution of electrons in molecules produces complicated and fluctuating changes in attraction and repulsion.

REMINDER

Dipoles are constantly being produced in molecules in solution

3.4 Hydrophobic interactions

Hydrophobic (meaning literally 'water hating') interactions differ from those intermolecular interactions already described in that they are not electrostatic in nature. Nor is there any attraction, or repulsion, between hydrophobic groups. The basis of hydrophobic interactions resides in the behaviour of water. Hydrophobic, or **apolar**, groups are so called because they are insoluble in water and do not interact with it. **Hydrophilic** ('water loving'), or **polar**, groups are soluble in water, and do interact with it. The **functional** groups that we have so far come across, such as carboxyl (-COOH), amine (-NH_2), hydroxyl (-OH), are all polar groups. They all contain an electronegative atom that will induce a dipole and an unequal distribution of charge. Water is a polar molecule; it will interact with other polar groups through electrostatic interactions. Molecules which contain polar groups are generally **soluble** in water.

On the other hand, apolar (hydrophobic) groups do not contain any particularly electronegative atoms. For example, the apolar methyl group (-CH_3) consists of carbon covalently bonded to three hydrogen atoms. The electronegativities of carbon and hydrogen are similar and there is therefore little tendency for electrons to assume an unequal distribution (no dipole is realised). The methyl group is therefore 'neutral'. Neutral groups have no way of interacting with water (no charge–charge interaction is possible). The water solvent will respond to such apolar groups by 'excluding' them. Apolar groups do not attract or repel each other, but rather coalesce (come together) through the action of the water solvent.

REMINDER

Apolar (hydrophobic) groups do not interact with water; they are 'neutral' groups in which the electronegativity of the constituent atoms is approximately equal

The 'hydrophobic effect' is of major importance in biology. Biological membranes form a hydrophobic barrier which defines a cell, or cellular organelle. Protein molecules fold and adopt specific three-dimensional structures, driven by the 'expulsion' of hydrophobic groups from water. The hydrophobic groups in proteins are found buried within the protein's interior, where there is no water.

TAKING IT FURTHER: Solubility in water (p. 42)

REMINDER

All biological molecules must interact (bind) in a short-lived, reversible manner to invoke a response. Such phenomena are mediated by intermolecular interactions

3.5 Summing up

1. Relative to covalent bonds, intermolecular interactions are weak, but collectively strong. They form the basis of recognition and binding between biological molecules.
2. Hydrogen bonds are a relatively strong form of intermolecular electrostatic interaction; they can form whenever a strongly electronegative atom (e.g. oxygen, nitrogen or fluorine) approaches a hydrogen atom which is covalently attached to a second strongly electronegative atom. Distance and direction are important in hydrogen bonds.
3. Charge–charge interactions (attraction or repulsion) occur between groups that carry a net electrical charge. Such interactions also occur at very small distances between molecules as a result of temporary, induced or permanent dipoles.
4. Hydrophobic, or apolar, interactions occur with groups that show no polarity (no unequal distribution of charge), and are therefore unable to interact with water. They are insoluble in water and are excluded from this solvent.

3.6 Test yourself

The answers are given on p. 179.

Question 3.1
(a) What type of intermolecular interaction is likely to occur when a strongly electronegative atom approaches a hydrogen atom that is covalently attached to a second electronegative atom?
(b) Give three examples of common biological functional groups which may participate in this type of intermolecular interaction.

Question 3.2
Give three characteristics of the hydrogen bond which distinguish it from other types of charge–charge interaction.

Question 3.3
What are the differences which distinguish intermolecular interactions from intramolecular (covalent) interactions?

Question 3.4
Induced or temporary dipoles in molecules are important in what type of intermolecular interaction?

Question 3.5
Which of the following groups are considered 'hydrophobic'?
(a) $-CH_2OH$
(b) $-SH$
(c) $-CH_2CH_3$
(d) $-CH_2NH_2$
(e) [benzene ring]
(f) [benzene ring]—OH

Taking it further

Solubility in water

We have discussed the fact that atoms and molecules are held together by various types of intermolecular interaction, including hydrogen bonds, charge–charge interactions and van der Waals forces. These forces are intricately involved in solubility because it is the solvent–solvent, solute–solute, and solvent–solute interactions that govern solubility.

Water is a polar molecule. As we have seen, its strongly electronegative oxygen atom induces a charge distribution in the molecule (a dipole). This gives rise to hydrogen bonding between water molecules; thus water is a very ordered structure!

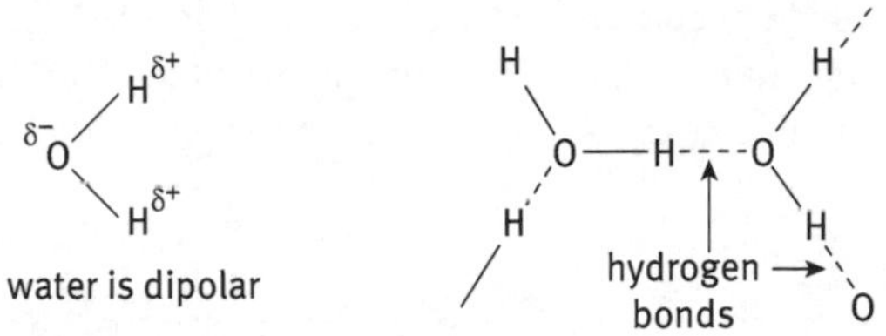

water is dipolar

hydrogen bonds

In order for substances to dissolve in water, room has to be made within this ordered structure to accommodate the substance.

If such a substance can interact with water molecules, through hydrogen bonding or charge–charge interactions, then it will be accommodated (it is a polar molecule) and consequently it will dissolve. By definition, a polar substance is one which interacts with water. Polar molecules are often referred to as hydrophilic, 'water loving'.

For simple substances such as salt, the small and charged Na^+ and Cl^- ions are readily accommodated within the water structure. Water molecules surround and interact with ions through charge–charge interactions, forming a 'cage' or hydration sphere to surround the ion (Fig. 31).

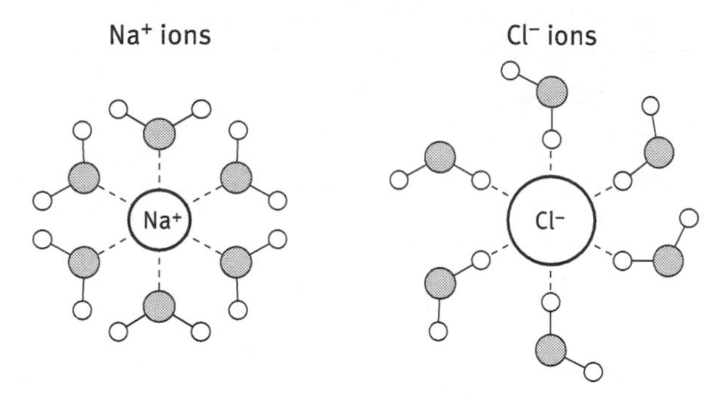

Figure 31. Water molecules surround ions in solution

For larger molecules which contain polar groups, the situation is somewhat more complex but the principle is the same. Biological molecules can contain a variety of polar functional groups, any of which will confer solubility in water. Hydroxyl groups (-OH), carboxyl groups (-COOH), carbonyl groups (-C=O), aldehydes (-CHO), sulphydryl groups (-SH) and amino groups (-NH_2) all contain an electronegative atom and all carry a charge under physiological conditions. As a general rule, the more polar groups on a molecule, the more soluble that molecule is.

Large, polar molecules, will naturally be surrounded by a relatively large hydration sphere (Fig. 32).

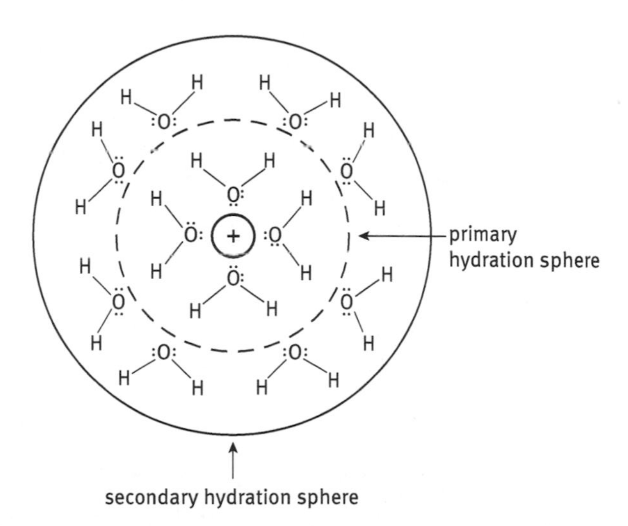

Figure 32. Hydration spheres surround larger polar molecules in solution

The hydroxyl group (-OH) on biomolecules is particularly important in conferring solubility through its ability to form hydrogen bonds (shown by the dashed lines in Fig. 33).

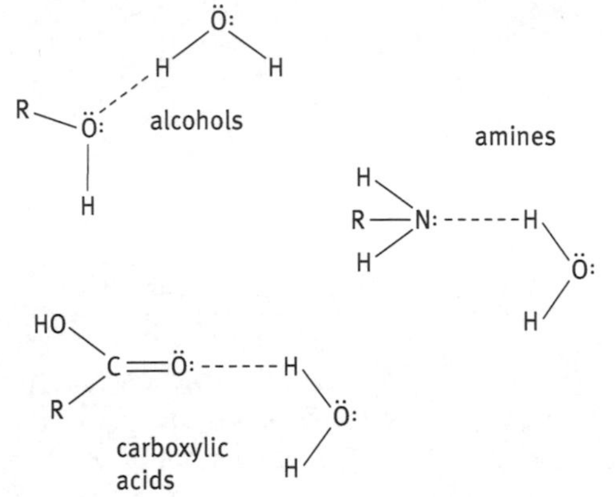

alcohols

amines

carboxylic acids

Figure 33. Hydroxyl groups readily partake in hydrogen bonding

As a general rule, the more polar groups a molecule has (and the smaller the molecule) the more soluble it is in water. Simple sugars such as glucose and fructose, with an abundance of hydroxyl groups, are very water soluble.

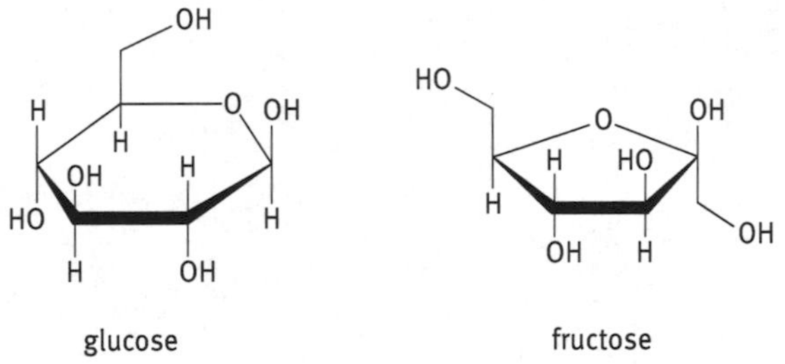

glucose

fructose

QUESTION

Look at the molecules listed in the table below. Which is the most polar?

A	CH_3-CH_2-OH
B	CH_3-CH_2-CH_2-CH_2-OH
C	CH_3-CH_2-CH_2-CH_2-CH_2-CH_2-OH
D	CH_3-CH_2-CH_2-CH_2-CH_2-CH_2-CH_2-CH_2-CH_2-CH_2-OH

> Answer: **A** is the most polar; it has a hydroxyl group and is a small molecule. The solubility of these molecules in water decreases from A to D. The methyl (-CH_3) and methylene (-CH_2) groups are both apolar; they are hydrophobic 'water hating' groups. There is no 'electronegative effect' in these groups, no opportunity for hydrogen bonding with water, no dipole and therefore no charge–charge interaction.

A major group of biological molecules is the lipids, which includes fats, oils and waxes. A major constituent of lipids are fatty acids. Fatty acids consist of a long hydrocarbon chain that is apolar (hydrophobic), which terminates in a polar (carboxyl) group. The molecule is said to be amphipathic; it has both a polar end and an apolar end (Fig. 34).

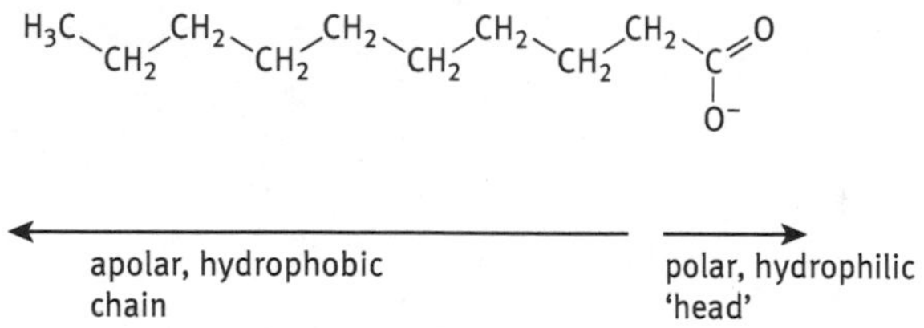

$H_3C-CH_2-CH_2-CH_2-CH_2-CH_2-CH_2-CH_2-CH_2-C(=O)O^-$

apolar, hydrophobic chain ← → polar, hydrophilic 'head'

Figure 34. An amphipathic fatty acid molecule

Fatty acids are linked together with glycerol and phosphate to form phospholipids, the major constituent of biological membranes. The highly polar phosphate group makes this a very amphipathic molecule (Fig. 35).

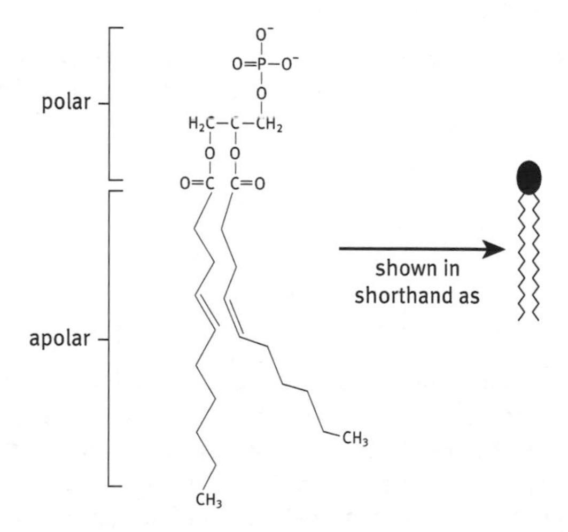

Figure 35. Amphipathic phospholipids are major constituents of biological membranes

In water these phospholipid molecules align themselves to form a 'bilayer', the basic structure of biological membranes (Fig. 36). Polar 'head groups' face and interact with the water environment, whereas the apolar hydrophobic fatty acid 'tails' form a hydrophobic environment from which water (and other polar molecules) are excluded.

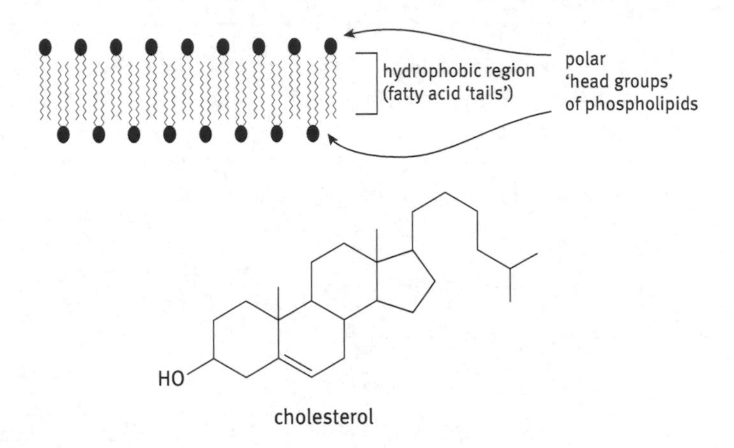

Figure 36. The basic structure of a biological membrane is a phospholipid bilayer

The steroid molecule cholesterol (Fig. 36) is a common constituent of the outer membranes of cells. The amphipathic character of this molecule can be appreciated through the hydroxyl group, which orientates this end of the molecule to the surface of the membrane in contact with water, and the hydrophobic fused ring structure and 'tail' which anchors the molecule within the hydrophobic interior of the cell membrane.

Although biological organisms are composed of greater than 60% by weight of water, they also contain an extensive network of biological membranes. In essence we have two completely opposite environments; the polar water environment and the apolar membrane environment. For molecules to move around the body they must traverse these two environments. Molecules will partition into these two environments dependent upon their polarity. This has enormous consequences for the distribution and pharmacological action of drugs. The science of pharmacology studies the absorption, metabolism, distribution and activity of drugs on the body.

As early as 1847, observations suggested that the more hydrophobic a substance, the more permeant it was; in other words, how more readily it was absorbed by the body, and these early observations remain valid today. It appears, however, that all hydrophilic molecules, and most hydrophobic molecules, require specific carriers (transporters) for their movements across cell membranes. Mammalian genomes probably contain in excess of 1000 genes encoding such transporters!

Taking it further

The gas laws

Intermolecular forces exist between all types of molecules and are responsible for holding them together in solids and liquids. In gases the molecules are moving more quickly and are further apart so the intermolecular forces are much weaker.

The **pressure** of a gas results from the molecules in the gas bombarding the sides of the container. The more collisions that occur, the greater the pressure of the gas.

Gas molecules are in constant and random motion. This motion is called **diffusion** and causes the molecules in a gas to spread apart and completely fill the container.

We can describe the physical properties of a gas by its pressure (P), volume (V), temperature (T) and quantity (or number of moles) (n). All these properties are related by a set of laws called the **gas laws**. Gases which obey these laws perfectly are called **ideal gases** but in practice real gases only approximately obey the laws.

Boyle's law

Boyle's law relates the pressure and volume of a certain amount of gas at a fixed temperature. Consider a sample of gas in a container such as a cylinder fitted with a piston (Fig. 37). The initial pressure of the gas is equal to P1 and its volume V1. If we increase the pressure on the gas to P2 by pushing the piston into the cylinder, the volume of the gas decreases to V2. Boyle's law states that the volume of a gas (V) is inversely proportional to its pressure (P). We can write this mathematically as:

$$V \propto 1/P$$

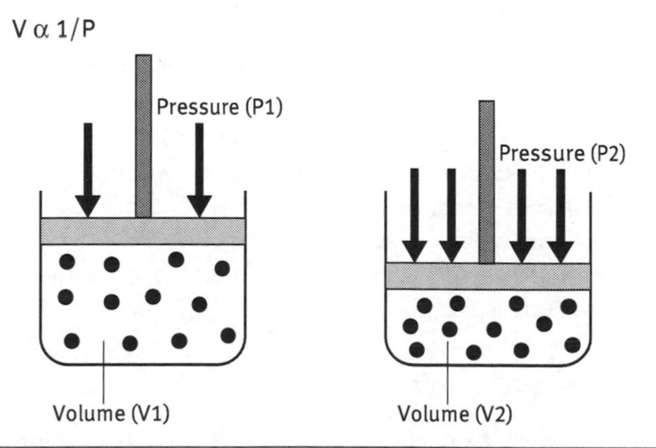

Figure 37. Boyle's law

Charles's law

This law relates the volume of a gas to its temperature. Consider a fixed mass of gas in the same cylinder fitted with a moveable piston with an initial volume V1 and at a temperature T1. If we heat the gas up to T2, the molecules will move more quickly and push out the piston until the pressure inside the cylinder is the same as the external pressure (Fig. 38). The volume of the gas increases as the piston is pushed back to the new volume V2.

This can be written as:

$$V \alpha T$$

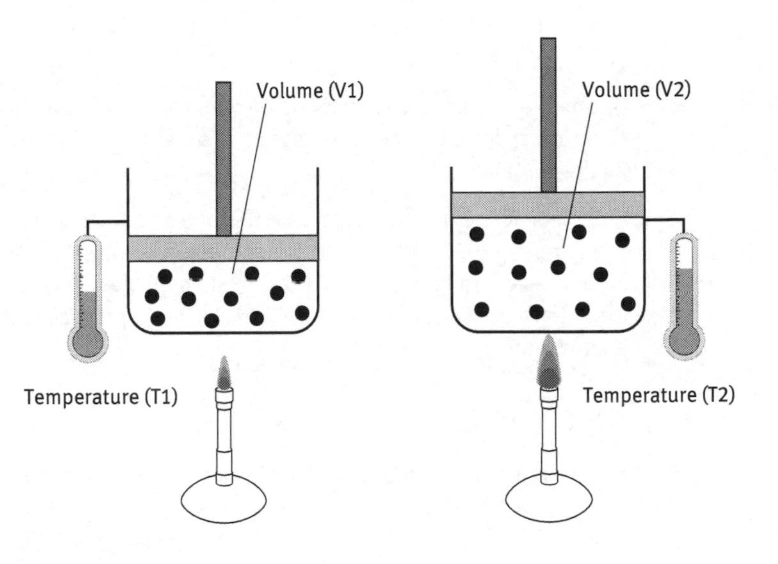

Figure 38. Charles's law

Gay-Lussac's law

There is a further law which relates the pressure of a gas to its temperature. Consider a gas in a cylinder with a volume of V1 at a temperature of T1. When the gas is heated to the new temperature T2, the kinetic energy of the molecules increases. However, if the volume of the cylinder is fixed and the piston cannot move then the number of collisions of molecules on the cylinder walls increases and so the pressure of the gas inside increases to a new pressure P2. This law states that for a gas with a constant volume the pressure is directly proportional to its temperature, $P \alpha T$.

Avogadro's law

One final law relates the volume of a gas to the quantity present and states that at a fixed temperature and pressure the volume of a gas (V) is directly proportional to the number of molecules (or moles) present (n). [You will meet the mole in the next chapter but at the moment it can be thought of as a way of measuring the amount of a substance.]

Mathematically this can be written as:

$$V \alpha n$$

From this it follows that equal volumes of gases at the same temperature and pressure contain the same number of molecules. In fact one mole of any gas occupies the same volume and at standard temperature and pressure (20°C and 1 atm pressure) this volume is 24.79 dm^3. This is known as the molar volume of a gas, V_m.

Molar volume, $V_m = 24.79\ dm^3$ at STP

The ideal gas law

Combining the four gas laws above gives us the ideal gas equation

$$PV = nRT$$

This equation relates the pressure (P), volume (V), temperature (T) and number of moles of a gas (n) and includes a constant, R, called the **universal gas constant.**

If the pressure of the gas is measured in pascals (Pa), the volume in m^3 and the temperature in kelvin (K), the value of the gas constant is 8.3145 J mol^{-1} K^{-1}.

The ideal gas law shows that the physical properties of pressure, volume, temperature and quantity of a gas are not independent. If three of these values are known, the fourth is fixed.

The equation is the same no matter which gas we use, but the law is only obeyed precisely for ideal gases.

A NOTE ON UNITS:

The SI unit of pressure is the pascal, Pa. The more familiar unit of pressure is the atmosphere (atm). 1 atm = 101 325 Pa or approximately 1×10^5 Pa.

The fundamental SI unit of length is the meter (m) and so volume is expressed in m^3.

One dm = 0.1 m, so 1 dm^3 (or 1 litre) = 10^{-3} m^3.

The SI unit of temperature is the kelvin (K). On the Kelvin scale 0 K is equivalent to -273°C and 0°C = 273 K. To convert from degrees Celsius to Kelvin we add 273. One Kelvin degree is the same size as one degree Celsius.

Partial pressure of a gas

If we have a mixture of gases in a container each individual gas exerts a pressure, known as the partial pressure of the gas. The partial pressure of each gas in a mixture depends upon the fraction of the gas present and the total pressure of the gas. So for a gas A in a mixture of gases A and B the partial pressure is given by:

$$p_A = P \times \chi_A$$

and the partial pressure of gas B by:

$$p_B = P \times \chi_B$$

where p = partial pressure of individual gas in mixture, P = total pressure of gas, χ = fraction of gas.

Air is a mixture composed mainly of nitrogen (about 78%) and oxygen (about 21%), with a small amount of carbon dioxide, water vapour and other trace gases. So for air at 1 atmosphere pressure the partial pressure of nitrogen, P_{N_2} is 1 atm x 0.78 = 0.78 atm and the partial pressure of the oxygen, P_{O_2}, = 1 atm x 0.21% = 0.21 atm.

The total pressure of a gas (P_T) is equal to the sum of the partial pressures of the gases in the mixture.

$$P_T = p_A + p_B$$

Solubility of gases in liquids and Henry's law

The amount of gas that will dissolve in a liquid depends largely on the pressure of the gas above the liquid. The greater the gas pressure, the more gas will dissolve in the liquid. Thus if we have a mixture of gases dissolving in a liquid, the greater the partial pressure of the gas, the more gas will dissolve in the liquid.

Henry's law states that the amount of gas dissolved in a liquid is proportional to the pressure of the gas above the liquid at a fixed temperature. The constant which relates the two quantities is Henry's law constant, $\boldsymbol{k_H}$. The value of k_H depends upon the gas, the solvent and the temperature:

$$\mathbf{c} = \boldsymbol{k_H} \times \mathbf{P}$$

where c = concentration of the gas in the liquid, k_H = Henry's law constant, P = pressure of the gas.

For a mixture of gases the solubility of each gas depends upon the partial pressure of the gas and the Henry's law constant for that gas:

$$c_A = k_{HA} \times p_A$$

Henry's law has serious implications for deep sea divers and explains the 'bends' or decompression sickness experienced on re-surfacing. At great depths under water, the pressure of air increases. As the pressure increases so does the solubility of air in blood on inhalation. Because nitrogen is the major component of air, large excesses of nitrogen build up in the blood. The dissolved oxygen is used in metabolism but the nitrogen is not. As the diver returns to the surface the pressure reduces and the nitrogen is released from the blood too quickly, forming bubbles. These result in severe pains in the joints, and can also affect the brain, skin, lungs and inner ear. In order to avoid decompression sickness, deep sea divers and similar workers are brought to surface pressure slowly, allowing the dissolved gases to be released at a rate that does no damage.

04 Counting molecules

BASIC CONCEPTS:

Atoms join together to make molecules, molecules join together to make larger molecules or react to form new molecules. Since molecules come in a variety of shapes and sizes, it is essential that we have some simple way of comparing and counting them. For example, we might need to make up a solution with a specific ratio of compounds at a certain concentration, or express the activity of an enzyme in terms of how many molecules of substrate it converts in a given time. We can compare the concentration of molecules in solution through the concept of the mole. This is a 'must know' concept in almost every branch of biology.

4.1 Moles

As biologists, we are interested in the interaction between molecules, so we need a system whereby we can compare the number of molecules of substances in solution. That system uses the **mole**. A mole is a certain number of molecules. The system works like this:

- Firstly we need a standard. The standard we use is the carbon-12 atom, $^{12}_{6}C$, the stable isotope of carbon containing 6 protons and 6 neutrons. A convention was adopted for expressing the atomic mass of an element in terms of atomic mass units (amu). This was defined such that the carbon-12 atom is exactly equal to 12.0 amu. Thus 1 amu is equal to one-twelfth the mass of an atom of carbon-12. 1 amu = 1.661×10^{-24} g.
- Every element in the periodic table can be assigned an atomic mass, which is its mass relative to that of one-twelfth the mass of an atom of carbon-12. This is known as the relative atomic mass (RAM) scale.
- Because the amu is very small it is more convenient to express the relative atomic mass in grams. We therefore define the quantity which is equivalent to 12g of carbon-12 to be equal to 1 mole.
- 1 mole, or 12 g of carbon-12, contains 6.022×10^{23} atoms of carbon-12.
- 1 mole, or the amount equal to the atomic mass in grams of any substance contains 6.022×10^{23} atoms of the substance.
- The number 6.022×10^{23} is known as Avogadro's number or Avogadro's constant.

One mole of a substance is the amount of substance equivalent to the formula mass in grams of the substance and contains Avogadro's number, 6.022×10^{23} atoms, molecules or ions of the substance.

From our definition, one mole of a compound has the identical number of mass components as there are atoms in 12 g of carbon-12. So, 12 grams of carbon is equivalent to 1 mole and contains 6.022×10^{23} atoms of carbon.

The mass of a water molecule (H_2O) is 18 amu (one oxygen =16 amu + two hydrogens = 2 amu). Therefore 18 grams of water is 1 mole and contains 6.022×10^{23} molecules of water.

Glucose ($C_6H_{12}O_6$) has a molecular mass of 180 amu (by adding the atomic mass units of all the atoms in a glucose molecule), so 180 grams of glucose is 1 mole (containing 6.022×10^{23} molecules of glucose).

One formula unit of sodium chloride, NaCl, has a mass of 58.45 amu. Therefore 1 mole of sodium chloride has a mass of 58.45 g. This mass of sodium chloride contains 6.022×10^{23} sodium (Na^+) ions and 6.022×10^{23} chloride (Cl^-) ions. This information is summarised in the table.

Formula	Formula mass of 1 unit	Formula mass of 1 mole	Number of single units in 1 mole	Number of atoms or ions in 1 mole
$^{12}_{6}C$	12 amu	12 g	6.022×10^{23} atoms C	6.022×10^{23} atoms C
H_2O	18 amu	18 g	6.022×10^{23} molecules H_2O	6.022×10^{23} atoms of O
				$2 \times 6.022 \times 10^{23}$ atoms of H
$C_6H_{12}O_6$ glucose	180 amu	180 g	6.022×10^{23} molecules glucose	$6 \times 6.022 \times 10^{23}$ atoms C
				$12 \times 6.022 \times 10^{23}$ atoms of H
				$6 \times 6.022 \times 10^{23}$ atoms of O
NaCl	58.45 amu	58.45 g	6.022×10^{23} units NaCl	6.022×10^{23} Na^+ ions
				6.022×10^{23} Cl^- ions

Because 12.0 g of carbon is an easily manageable amount of substance to use, the mole became the standard unit for counting numbers of atoms or molecules.

The MOLE is a number just like a dozen or a century! There are exactly the same number of molecules in a mole of glucose as there are in a mole of insulin. We can therefore compare the numbers of molecules of different substances in solution directly, irrespective of whether they are small or large molecules. For example, most enzymes are very large molecules compared to their substrates. Nevertheless, one enzyme molecule reacts with one substrate molecule at a time, so it is important to be able to work out how many molecules of each you might have in solution.

QUESTION

How many moles are there in 57.5 g of sodium?

Sodium has 23 atomic mass units, so 23 g of sodium would be equal to 1 mole. So, in 57.5 g sodium, there must be 57.5/23 = 2.5 moles

4.2 Molecular mass

We have seen that the mass of one mole of any compound can be obtained by summing the relative atomic masses of all the elements which make up the compound and converting to grams. The mass of one mole of a compound in grams is called its **molar mass, *M***.

So we have the relationship:

$$\text{Number of moles} = \frac{\text{mass in grams}}{\text{molar mass}} = \frac{m}{M}$$

A variety of symbols are used to describe the molecular mass of compounds.

- The molar mass (symbol M) of glucose is 180 g mol^{-1}. This is the mass of one mole.
- The relative molecular mass (symbol M_r) of glucose is 180. This is the relative mass of one molecule compared to one-twelfth the mass of one atom of carbon-12.
- The name 'dalton' (symbol Da), is used by biologists as an alternative to the cumbersome name of 'atomic mass unit' for one-twelfth of the mass of the atom of carbon-12. Thus, the molecular mass of glucose can be expressed as 180 Da.

These are all correct ways of saying the same thing by present recommendations. Note that M_r does not have units; this is because relative molecular mass is the ratio between the mass of a molecule and the mass of one-twelfth of a carbon atom, and as a ratio it has no units.

4.3 Moles and molarity

A **mole** refers to the amount of a substance whereas the **molarity** refers to its concentration.

Moles and molarities are a common source of confusion amongst students, partly because of the similarities in their names, so it's important to make sure that you know precisely what these terms mean.

REMINDER

A mole is a number, an amount, whereas molarity is a concentration

A **mole** (abbreviation mol) is a measure of the amount of a substance.

A mole of a compound is the amount which contains a number of molecules equal to Avogadro's number (6.022×10^{23}). Or a mole of compound is the amount of a substance equal to the relative molecular mass expressed in grams.

Molarity is a measure of concentration. The molarity of a solution measures the number of moles of a substance in a certain volume.

REMINDER

Avogadro's number is equal to the number of molecules in a mole ($= 6.022 \times 10^{23}$)

Now, in terms of numbers of molecules, it does not matter if you have 1 mole of a substance in 1 litre, or in 1 ml of solution, you still have the same number of molecules (because 1 mole = 6.022×10^{23} molecules). However, the concentration must change; 1 mole of substance in 1 ml is clearly a more concentrated solution than 1 mole in 1 litre. To define concentration, we refer to molarity.

The molarity of a solution is the number of moles per litre of solution (or number of moles per dm^3, if you prefer). The units of molarity are mol/l or mol l^{-1} or M.

A molar solution (abbreviation 1 M) contains one mole of a material in a volume of one litre.

So, given the M_r of glucose as 180, then 180 grams of glucose is 1 mole, and 180 grams of glucose dissolved in 1 litre of water would give a 1 molar (1 M) solution.

If we take just 1 ml of this solution, then the concentration of glucose in that 1 ml is still 1 molar (we have not changed the ratio of the mass of glucose to the volume of water). However, 1 ml of the solution contains 1/1000 the number of moles (1 mmol). This is because 1 ml of solution is 1/1000 of a litre.

So a 3.2 M solution has a concentration of 3.2 mol/l or 3.2 mol l^{-1}. The mass or amount of the compound in the solution depends on the volume that you use. 1 ml of this solution contains only 3.2 mmol, or .0032 mol, (i.e. one thousandth of the amount contained in a litre) but the concentration remains the same.

REMINDER

$$\text{Concentration} = \frac{\text{number of moles}}{\text{volume (l)}} = \frac{n}{V} = c\ (\text{mol l}^{-1})$$

QUESTION

What is the molarity of water?

The M_r of water (H_2O) is 18, so 18 grams of water is 1 mole. One litre of water has a mass of 1000 grams, so it must contain

$$\frac{1000 \text{ g}}{18 \text{ g mol}^{-1}} = 55.5 \text{ moles of water, thus}$$

the molarity of water is 55.5 mol l^{-1} (55.5 M)

4.4 A note on units

Biologists frequently deal with solutions that are much less than 1 molar concentration. The terms milli (one thousandth), micro (one millionth) and nano (one thousand millionth) are commonly used to express concentrations, amounts or volumes of substances. These are abbreviated to 'm' (milli), 'μ' (micro) and 'n' (nano).

milli = 10^{-3} = m
micro = 10^{-6} = μ
nano = 10^{-9} = n

For example, if we took 1 ml (1 millilitre = one thousandth of a litre) of a 1 molar solution of glucose (containing 1 mole of glucose per litre), then that would contain only 1 mmol of glucose (one thousandth of a mole of glucose).

When tackling calculations involving moles or molarity, work from first principles. More examples are given in 'Taking it further: Confidence with moles'.

QUESTION

If we took 0.0105 g (10.5 mg) of a 10500 M_r protein, and dissolved this in 1 ml, then that 1 ml would contain 1 micromole (1 μmol) of the protein.

Why? because

$$\text{Number of moles} = \frac{\text{mass in grams}}{\text{molar mass}}$$

$$= \frac{0.0105 \text{ g}}{10\,500 \text{ g mol}^{-1}} = 0.000001 \text{ mol} = 1 \times 10^{-6} \text{ mol}$$

or 1 micromole (1 μmol)

TAKING IT FURTHER: Confidence with moles (p. 61)

4.5 Dilutions

Equally important as understanding moles and molarity, is understanding dilutions, since solutions are often diluted to achieve the required molarity! There are basically two ways of diluting solutions; simple dilutions and serial dilutions.

1. **Simple dilution:** A simple dilution is one in which a unit volume of a liquid material of interest is combined with an appropriate volume of a solvent liquid to achieve the desired concentration. For example, one litre of a solution of glucose in water is combined with a further litre of water to give a glucose solution of half the original concentration. The dilution factor is the total number of unit volumes in which your material will be dissolved. In the case above, the dilution factor would be 1:2 (verbalise as '1 to 2' dilution). In a further example, a 1:5 dilution entails combining 1 unit volume of diluent (the material to be diluted) + 4 unit volumes of the solvent medium (hence, $1 + 4 = 5$ = dilution factor). If you have a known volume and concentration of a substance, and you need to dilute this to give a different concentration, then the expression V1C1 = V2C2 is useful. V1 and C1 refer respectively to the original volume and concentration, and V2 and C2 to the desired volume and concentration.
2. **Serial dilution**: A serial dilution is simply a series of simple dilutions, which amplifies the dilution factor quickly, beginning with a small initial quantity of material. The source of dilution material for each step comes from the diluted material of the previous step. In a serial dilution the total dilution factor at any point is the product of the individual dilution factors in each step up to it. For example, say we start with 1 ml of a protein solution. If we take 0.1 ml (100 μl) of this, and add it to 0.9 ml of solvent, then we have made a 0.1:1 dilution (a 10-fold dilution). A further 0.1 ml from this, added to 0.9 ml solvent, gives another 10-fold dilution. The total dilution of the original protein solution is now $10 \times 10 = 100$. This is often the preferred way of making dilutions since it involves weighing out your substance only once, from which you can make a series of different concentrations. But remember, errors from dilutions can occur at every step and so the more dilutions that are made, the more errors that are possible.

More examples are given in 'Taking it further: Confidence with moles'.

4.6 Percent composition solutions

Sometimes, and usually out of convenience, solutions are made up as percent solutions.

REMINDER

Percentage (%) means number of parts per hundred,
i.e. 1% = 1 part per 100 total

1. **w/w percentage composition**: When both the solute (the substance being dissolved) and the solvent (the liquid doing the dissolving) are given in mass units (for example in grams), the percentage composition is written as % w/w. For example 20 g glucose dissolved in 480 g water has a % w/w of 2%. This means that by mass the glucose forms 2% of the total solution.

   Why? Because
   Mass solute (glucose) = 20 g
   Mass solvent (water) = 480 g
   Mass solution = 20 g + 480 g = 500 g
   % w/w = 20/500 × 100 = 2%

2. **w/v percentage composition**: This is the most common way of making up a solution in the laboratory. A certain mass of dry solute is weighed and placed in a container calibrated to contain a fixed volume. Solvent is added until the known, calibrated volume is reached. The concentration is then expressed in terms of percentage weight per volume of solution (% w/v). For example 10 g sodium chloride is dissolved in water to give 100 ml solution. The % w/v is therefore 10% w/v.

   Why? Because
   Mass solute (sodium chloride) = 10 g
   Volume of solution (sodium chloride and water) = 100 ml
   % w/v = (10/100) × 100 = 10%

REMINDER

A liquid changes volume only slightly when a solute is dissolved in it, whereas its mass changes significantly

3. **v/v percentage composition**: When using liquid reagents, the percent concentration is based upon volume per volume, i.e. volume of liquid solute per volume of solution. So 10 ml of liquid substance added to 90 ml of buffer would give a 10% v/v solution. If you wanted to make 70% ethanol you would mix 70 ml of 100% ethanol with 30 ml water.

   Why? Because
   70% v/v ethanol = 70 ml ethanol in 100 ml total solution
   The volume of water solvent is therefore 100 ml – 70 ml = 30 ml.

4.7 Summing up

1. **Molecular mass**
   The mass of a molecule is the sum of the masses of the atoms from which it is made. The unit of mass – the atomic mass unit (amu or u), the dalton (symbol Da) – all mean the same thing and are defined as one-twelfth of the mass of an atom of carbon-12.

2. **Relative molecular mass**
   This is the relative mass of one mole (or one molecule) compared to one-twelfth of the mass of one mole (or one atom) of carbon-12.

3. **Mole**
   A mole is the quantity of a substance whose mass in grams is equal to the molecular mass of the substance. One mole of substance contains 6.022×10^{23} particles of the substance.

4. **Molarity**
   Molarity is a measure of the concentration of a solution. If you dissolve 1 mole of a substance in 1 litre of solution, you have made a 1 molar (1 M) solution.

5. **Dilutions**
   For simple dilutions, the dilution factor is the total number of unit volumes in which your material will be dissolved; serial dilutions are a combination of simple dilutions that produce a final dilution factor equal to the product of the individual dilution factors.

6. **Percent solutions**
   For a solid, $\dfrac{\text{mass of solute}}{\text{volume of solution}} \times 100 = \text{percent concentration (w/v)}$.

   For a liquid, $\dfrac{\text{volume of solute}}{\text{total volume of solution}} \times 100 = \text{percent concentration (v/v)}$.

4.8 Test yourself

The answers are given on p. 179–180.

Question 4.1
If a solution contains 0.01 grams of insulin, and the M_r of insulin is 6000, how many moles of insulin are present?

Question 4.2
How many grams of ethanoic acid ($C_2H_4O_2$) would you need to make 10 litres of a 0.1 M ethanoic acid solution? (Take the relative atomic masses of carbon as 12, oxygen as 16 and hydrogen as 1)

Question 4.3
50 ml of a glucose solution was prepared by dissolving 10 grams of glucose in 50 ml of water. Given that the relative molecular mass (M_r) of glucose is 180, how many moles of glucose are present in this solution, and what is the molarity of the solution?

Question 4.4
A stock solution of the amino acid glycine was 0.02 M. 1 ml of this solution was used in an enzyme assay, in a total volume of 3 ml. How many moles of glycine are present in the enzyme assay, and what is the molarity of glycine in the assay?

Question 4.5
1.2 grams of glycine ($M_r = 79$) was dissolved in 100 ml of water. In 1 ml of this solution, (a) how many moles of glycine are present, (b) what is the molarity of the glycine solution, (c) how many molecules of glycine are present?

Taking it further

Confidence with moles

Here are some examples in working out molarities, concentrations and dilutions.

Moles and molarity

A. To prepare a litre of a simple molar solution from a dry reagent.

Multiply the molecular mass by the desired molarity to determine how many grams of reagent to use:

If $M_r = 194.3$, and you need to make a 0.15 M solution, then use

$194.3 \times 0.15 = 29.145$ g/l

If you only need 30 ml of the above solution, then use $194.3 \times 0.15 \times 30/1000 = 0.87$ g

B. You take 50 μl of a 1 mM stock sugar solution and add it to the enzyme reaction mixture to give a final volume of 3 ml; what is the concentration of sugar in the reaction mixture?

50 μl (= 0.05 ml) into 3 ml is a dilution factor of $3/0.05 = 60$, so the final concentration of the sugar is $1/60 = 0.017$ mM (17 μM).

C. You take 50 μl of a stock sugar solution which contains 100 μmol of sugar per ml, and add this to 3 ml of enzyme reaction mixture. How many micromoles of sugar are there in the reaction mixture?

$0.05 \times 100 = 5$ μmol. This is what you have actually taken from the stock solution.

Adding this to the reaction mixture means you have 5 μmol in 3 ml, or 5/3 (= 1.66) μmol/ml.

D. You are given 10 ml of a stock solution of cytochrome *c* of 1 mg/ml and need to make 5 ml of a cytochrome *c* solution of 5 μg/ml.

(a) In other words, what volume of the stock solution do you need to add to a total volume of 5 ml to give 5 μg/ml?

In terms of V1C1 = V2C2, V1 is what we wish to know, C1 is 1 (mg/ml), V2 will be 5 (ml) and C2 needs to be 0.005 mg (=5 μg).

So, $V1 \times 1 = 5 \times 0.005$

From which V1 = 0.025 ml (or 25 μl). Therefore you would add 0.025 ml of the stock cytochrome *c* solution to 4.975 ml of water (or buffer) to give a final volume of 5 ml.

(b) Given that M_r for cytochrome *c* is 12 000, how many moles of cytochrome *c* are present in 1 ml of your new solution?

12 000 grams of cytochrome *c* would be equivalent to 1 mole. In 1 ml of solution you have 5 μg. Obviously the answer is going to be quite a small number; expressing in grams, 0.000005/12000 = 0.00000000042 moles, or 0.00042 μmol, or 0.42 nmol (nanomole, 10^{-9} of a mole).

(c) That's how many moles in 1 ml, so what is the molar concentration of our new cytochrome *c* solution?

Remember, molarity is expressed per litre. If we have 0.42 nmol in 1 ml, then we would have 0.42 × 1000 nmol in a litre (= 420 nmol). So this solution is 420 nM (nanomolar), or 0.420 μM, or 4.2×10^{-7} M

Dilutions

A. To convert from % w/v solution to molarity, multiply the percent solution value by 10 to get grams/l, then divide by M_r, the molecular mass.

$$\text{Molarity} = \frac{\text{M(\% solution)} \times 10}{M_r}$$

e.g. Convert a 6.5% solution of a substance with $M_r = 325.6$ to molarity,

[(6.5 g/100 ml) × 10]/325.6 g/l = 0.1996 M

B. To convert from molarity to percent solution, multiply the molarity by the M_r and divide by 10:

$$\text{\% solution} = \frac{\text{molarity} \times M_r}{10}$$

e.g. Convert a 0.0045 M solution of a substance having M_r 178.7 to a percent solution:

[0.0045 moles/l × 178.7 g/mole]/10 = 0.08% solution

C. You are given 5 ml of a 10% solution of glucose ($M_r = 180$). What is the molar concentration of your solution?

A 10% solution of glucose is equivalent to 10 g/100 ml; you have only 5 ml which must contain 10 × 5/100 = 0.5 g

If 5 ml contains 0.5 g glucose, a litre would contain 0.5 × 1000/5 = 100 g

180 g of glucose in 1 litre would be 1 M, so 100 g is equivalent to $1 \times 100/180 = 0.55$ M

And how many moles of glucose are in 5 ml?

$0.55 \times 5/1000 = 0.00275$ moles, or 2.75 mmol (millimoles, 10^{-3} of a mole).

D. Look at the table below. You are given a stock protein concentration of 1.5 mg/ml and need to make a set of simple dilutions to give a range of protein concentrations, in a fixed final volume of 3 ml. The figures have been included for you.

Stock (ml)	Water (ml)	Total volume (ml)	Concentration of protein (mg/ml)	Amount of protein (mg)
3	0	3	1.5	4.5
2.5	0.5	3	1.25	3.75
2	1	3	1.00	3
1.5	1.5	3	0.75	2.25
1	2	3	0.50	1.5
0.5	2.5	3	0.25	0.75
0	3	3	0	0

Firstly working out a protein concentration in mg/ml (second column from the right of the table), each is calculated by taking a dilution factor, which depends upon the final volume (always 3 ml). So, for example, 2 ml of stock into a final volume of 3 ml is a dilution of 2/3, so $2/3 \times 1.5 = 1.00$ mg/ml, and 0.5 ml stock into a final volume of 3 ml is a dilution of 0.5/3, so $0.5/3 \times 1.5 = 0.25$ mg/ml, and so on. The final figure is a concentration, amount per volume.

Very often we wish to know only the total amount of protein present, rather than its concentration. This is given in the last column of the table. This is calculated simply by multiplying the volume of stock used (in ml) by the amount of protein present in 1 ml (= 1.5 mg). The final figure for the amount of protein is independent of the final volume of liquid; for example, in this case it is 3 ml but if it were 30 ml it would not change the amount of protein present in each case. **It would, however, change the concentration (mg/ml) of protein present.** For example, in the third row of the table, 2 ml stock into a final volume of 30 ml would be a dilution of 2/30, so $2/30 \times 1.5 = 0.10$ mg/ml, 10 times less than the original, which is not surprising considering the volume has been increased 10 times!

Always be sure you know what you are dealing with, or what you are trying to calculate; i.e. is it an amount (grams, micrograms, moles) or is it a concentration (g/ml, μg/ml, molar)?

E. In a microbiology lab students perform a three-step 1:100 serial dilution of a bacterial culture (see diagram below). The initial step combines 1 unit volume culture (10 μl) with 99 unit volumes of broth (990 μl) = 1:100 dilution. In the next step, one unit volume of the 1:100 dilution is combined with 99 unit volumes of broth, now yielding a total dilution of 1:100 × 100 = 1:10 000 dilution. Repeated again (the third step) the total dilution would be 1:100 × 10 000 = 1:1 000 000 total dilution. The concentration of bacteria is now one million times less than in the original sample.

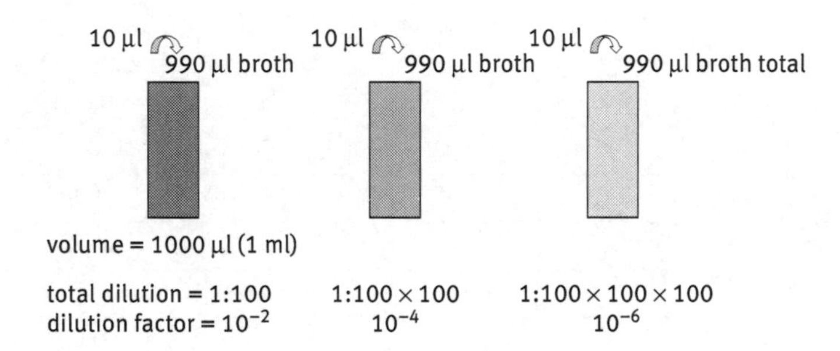

05 Carbon – the basis of biological life

BASIC CONCEPTS:
The element carbon is a central building block in life forms. Here we relate the electronic structure of carbon to its special properties, which are evident in the types of covalent compounds that it forms. Such properties determine the types and shapes of biological molecules. Carbon forms the skeleton and backbone of biomolecules.

All biological molecules contain carbon; indeed, biological life is based upon carbon. To understand why and how carbon fulfils this central role, we need to look at its bonding behaviour.

5.1 The electronic structure of carbon

Carbon atoms have six electrons. Two of them will be found in the 1s atomic orbital close to the nucleus. The next two will go into the 2s orbital. The remainder will be in two separate 2p orbitals. This is because the p orbitals all have the same energy and the electrons prefer to be on their own if possible (Fig. 39). The electronic structure of carbon is normally written $1s^2 2s^2 2p^2$.

This configuration suggests that carbon has **two unpaired** electrons in its outer p orbitals that could participate in the formation of two covalent bonds. However, we know from observation that this is not so. The molecule CH_2 does not exist. The simplest compound of carbon is methane, CH_4, which has four equivalent covalent carbon–hydrogen bonds.

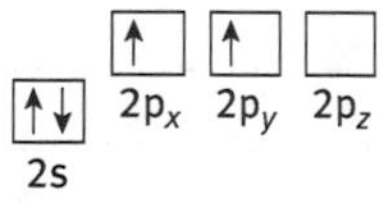

2s	$2p_x$	$2p_y$	$2p_z$
↑↓	↑	↑	

Figure 39. Outer electronic configuration of carbon

5.2 Hybridisation

To explain how it is possible for carbon to form four equivalent covalent bonds we use a modified theory of bonding called **orbital hybridisation.** **Orbital hybridisation** proposes that, prior to the formation of covalent bonds, carbon undergoes a change to its electron configuration (Fig. 40).

An electron from the 2s atomic orbital is elevated within energy level 2 to the empty 2p atomic orbitals (this requires a small input of energy). The four orbitals mix or hybridise to produce four equivalent hybrid orbitals. The electrons redistribute so that each of the now equivalent four hybrid atomic orbitals contains one electron. The new hybrid orbitals are called sp^3 orbitals. This then provides carbon with a maximum of four unpaired electrons that could form covalent bonds. In methane each half-filled sp^3 orbital overlaps with a hydrogen 1s orbital (containing one electron) to form a single covalent C-H bond.

Thus in methane, the carbon is said to be **sp^3 hybridised**; in other words, all four of the resulting hybrid orbitals are used in bonding. These have resulted from a 2s orbital and three 2p orbitals, hence sp^3.

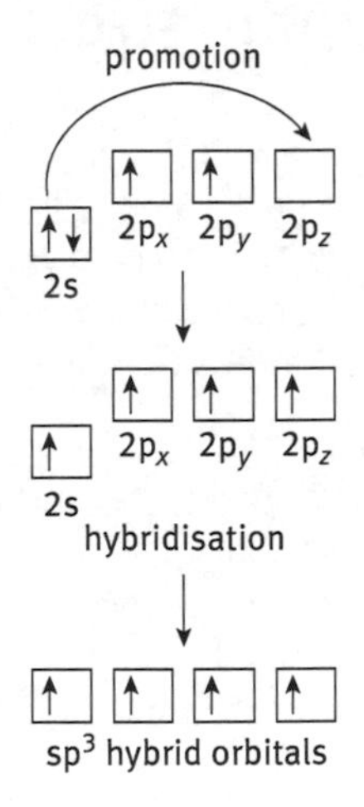

Figure 40. Hybridisation in carbon

TAKING IT FURTHER:
The peptide bond
(p. 28)

REMINDER

Orbital hybridisation occurs when atomic orbitals in an energy level mix, or hybridise, to form hybrid orbitals of the same energy

5.3 The tetravalency of carbon

When covalent bonds are formed, energy is released. The lower the energy content of a molecule, the more stable it is. Atoms will always attempt to make as many covalent bonds as possible to attain greater stability. Carbon is particularly good at this since it can form four covalent bonds. Carbon is said to be **tetravalent**. The tetravalency of carbon is central to its versatility as a biological building block. Carbon–carbon covalent bonds form readily, giving rise to both ring structures and chain compounds.

Every carbon atom in these rings has made four covalent bonds.

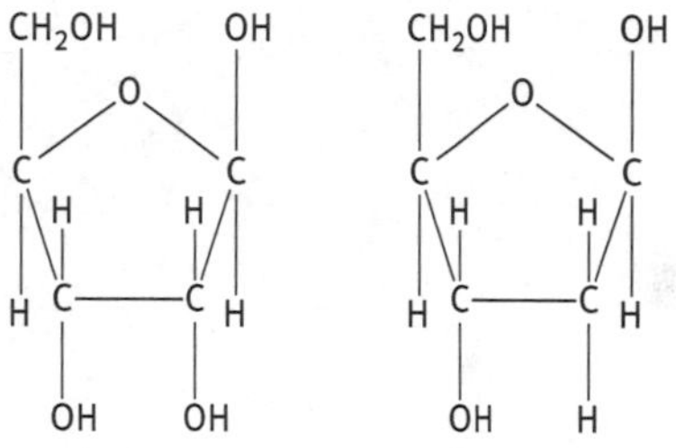

ribose

deoxyribose

The presence of carbon in ring structures is generally taken for granted (as indeed is the presence of hydrogen atoms attached to the carbon atoms), and so we would show deoxyribose, for example, in shorthand as

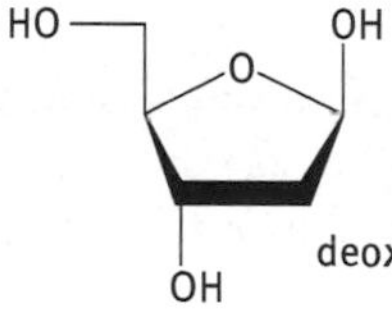

deoxyribose

Only the functional groups are shown, in this case the hydroxyl groups.

Long-chain structures are possible, such as those in the fatty acids. Again, each carbon in this palmitic acid chain is tetravalent.

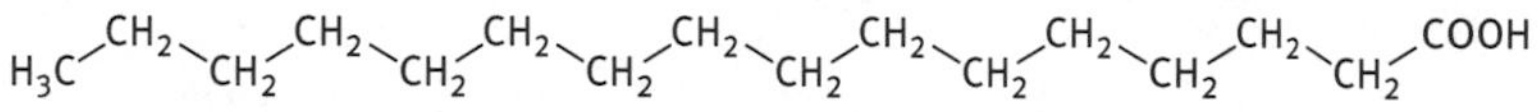

$H_3C-CH_2-CH_2-CH_2-CH_2-CH_2-CH_2-CH_2-CH_2-CH_2-CH_2-CH_2-CH_2-CH_2-CH_2-COOH$

palmitic acid

5.4 Shapes of molecules

When we draw structures such as those above, we are confined to a two-dimensional medium. Of course, many molecules are not flat, but rather three-dimensional. The shape of biological molecules is heavily dependent on the tetravalency of carbon. In an sp^3 hybridised carbon atom, the four hybrid atomic orbitals arrange themselves to be as far apart from each other as possible, each pointing to the corners of a regular tetrahedron. The shape of methane (Fig. 41) is therefore a tetrahedral structure.

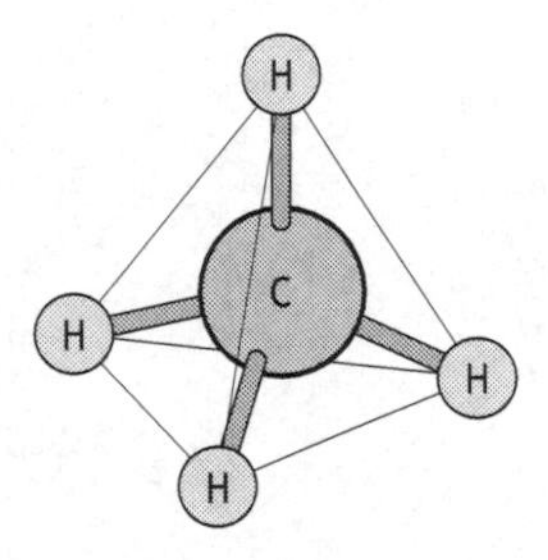

Figure 41. Methane has a tetrahedral structure

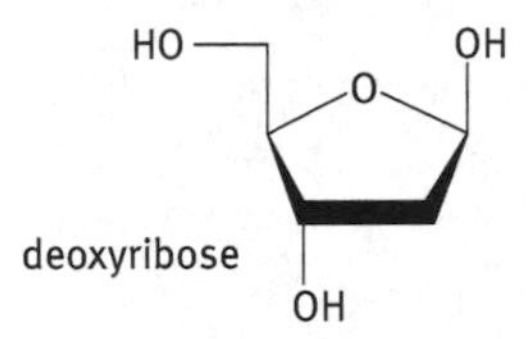

In this shorthand depiction of deoxyribose, the lower half of the ring is shown as a bold line; we use this to indicate that this part of the ring is coming out of the page. In other words, the ring is not flat; it has a three-dimensional shape. That shape is a consequence of the tetravalency of carbon.

In a similar way, we use different depictions to show the directions of covalently bonded groups in a molecule.

A normal bond is in the plane of the page; a wedge bond is coming out of the plane towards us, a hatched bond going into the plane away from us.

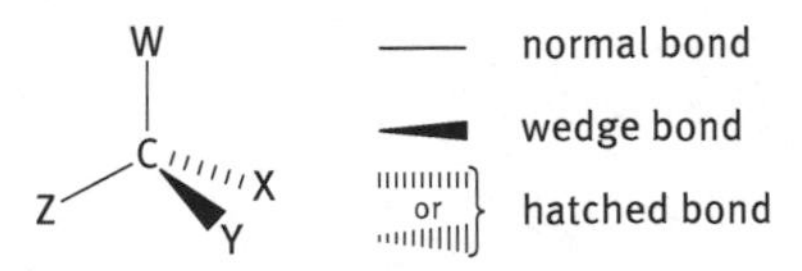

Ring structures in which carbon is bonded to four other atoms are not flat, but rather adopt 'chair' or 'boat' conformations. This is because the geometry at each sp^3 hybridised carbon atom is tetrahedral.

REMINDER

sp^3 hybridisation means that carbon can form four covalent bonds

For glucose the chair structure is the most stable; in the boat form there is steric crowding, with the hydroxyl groups being pushed together.

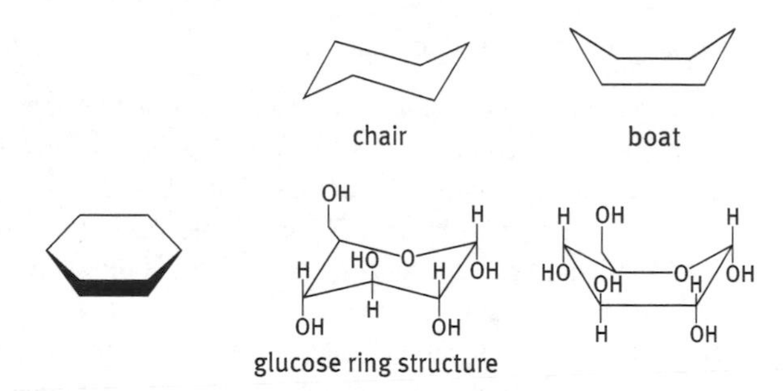

glucose ring structure

With larger molecules, many shapes or conformations are possible. Whereas carbon makes this possible, the ultimate three-dimensional shape of such large molecules is determined by the variety of **intermolecular interactions** between **functional groups** on the molecule, which act to constrain and hold the molecule in a particular three-dimensional conformation. The structural and functional properties of large biological macromolecules are determined by their three-dimensional shapes.

Molecular shape is the determining factor in binding and recognition events in biology. The active site of an enzyme must be able to accommodate the shape of its substrate, which in turn imparts a high degree of specificity for the interaction. Similarly, membrane receptors will recognise a particular hormone, or membrane transporters will move only a particular molecule. With shape comes specificity, and with specificity comes recognition, control and order; the hallmarks of a biological system.

5.5 Carbon in chains and rings – delocalisation of electrons

In those carbon ring structures shown above, each carbon atom has adopted an sp^3 hybridisation and forms four covalent bonds. However, carbon is capable of other types of hybridisation.

In compounds of carbon which contain double bonds, sp^2 hybridisation takes place. The sp^2 orbitals are formed in the same way as sp^3 orbitals but only two of the p orbitals are hybridised, leaving one unchanged p orbital.

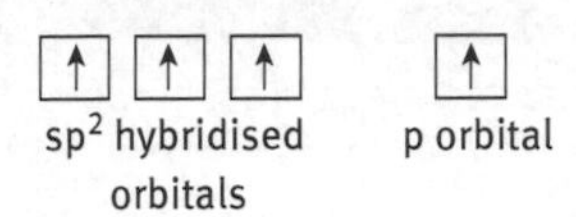

The three new sp^2 hybrid orbitals arrange themselves as far apart as possible and lie in a flat plane, 120° apart. The unchanged p orbital is perpendicular to the plane of the sp^2 orbitals.

This type of bonding occurs in ethene, C_2H_4, which possesses a double bond between the two carbon atoms and four single C-H bonds (Fig. 42). Each of the carbon atoms has three sp^2 hybridised orbitals and one p orbital. Each of the orbitals contains one electron. One of the electrons in the sp^2 orbitals overlaps ('head-to-head') with the electron from the sp^2 orbital on the second carbon atom to form a carbon–carbon sigma bond. The other two sp^2 orbitals on each carbon atom overlap with hydrogen 1s orbitals to give single C-H bonds. All six atoms are in the same plane; ethene is a planar (flat) molecule. The unchanged 2p orbitals on each C atom overlap sideways on to give a pi bond between the atoms. The pi bond is weaker than the sigma bond because only a small degree of overlap is possible.

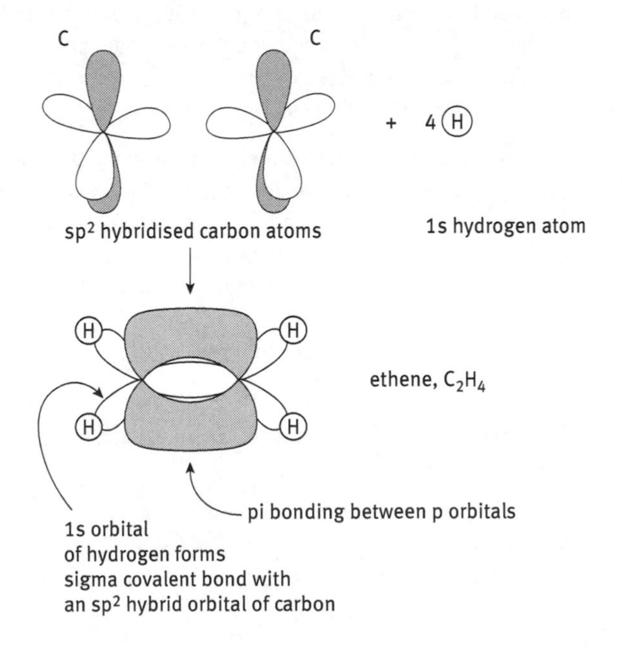

Figure 42. sp^2 hybridisation and bonding in ethene

The classic example of sp^2 hybridisation in carbon is shown in the molecule benzene, C_6H_6 (Fig. 43).

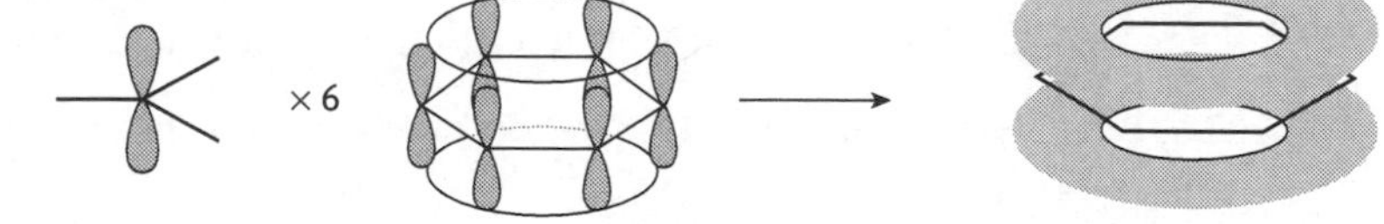

unused p orbital

$\times 6$

sp^2 hybridised C

delocalised π system

delocalised electrons in the benzene ring

Figure 43. sp^2 hybridisation and delocalisation of electrons in benzene

The benzene molecule consists of a ring of six carbon atoms, each attached to one hydrogen atom. Each carbon atom in benzene is sp^2 hybridised. The carbon uses its sp^2 orbitals to form three sigma bonds, one to each of the neighbouring carbon atoms and one to a hydrogen atom. The six carbon atoms and six hydrogen atoms are in the same plane. The six p orbitals, which are above and below the plane, are close enough to each other to overlap and form three pi bonds. This arrangement results in a continuous circular overlap of the six p orbitals. The electrons from each p orbital are free to move within this circular overlap region and are said to be **delocalised**.

REMINDER

In sp^2 hybridisation, carbon forms three equivalent hybrid orbitals, but one p orbital remains unchanged. This unchanged p orbital is frequently able to participate in pi bonding

The shorthand structures shown below are all correct ways of depicting the benzene molecule.

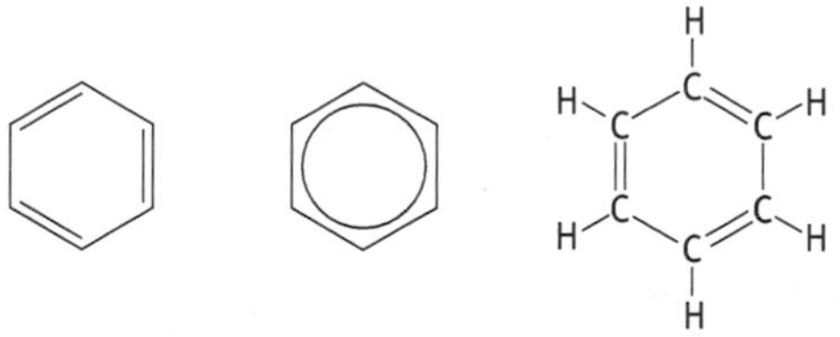

5.6 Aromaticity

With electron delocalisation comes increased stability; benzene is a relatively unreactive compound, and the ring conformation is planar. Benzene is an aromatic compound. A molecule is aromatic, i.e. it displays **aromaticity,** if

- it is fully **conjugated** (i.e. there is a 2p orbital on every atom in the ring)
- it is cyclic
- it is planar
- it contains 6, 10, 14 (i.e. $4n + 2$, where n is any integer) etc. delocalised electrons (Huckel's rule).

Ring structures (and chain structures) may be conjugated without being aromatic. A conjugated system may be represented as a system of alternating single and double bonds.

$$-C=C-C=C-C=C-$$

In such systems conjugation is the interaction of one p orbital with another across an intervening sigma bond. The ring structure (a porphyrin ring) in the molecule haem (Fig. 44), is a conjugated ring structure; there is a system of alternating single and double carbon bonds around the full extent of the ring.

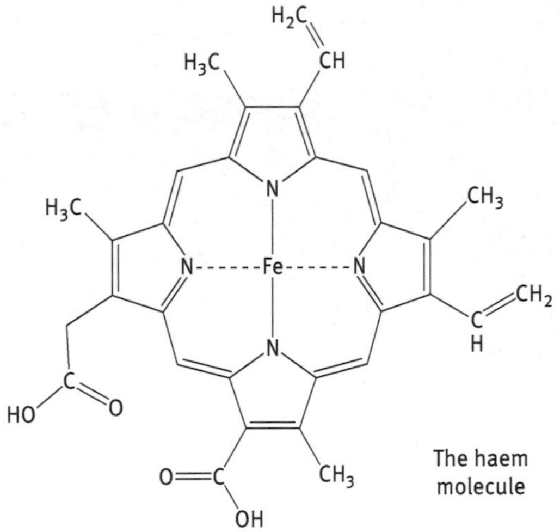

The haem molecule

Figure 44. A conjugated porphyrin ring

5.7 Functional groups and carbon families

We have seen that carbon can bond in a variety of different ways to other atoms of carbon and to hydrogen. However, it is when carbon bonds to different atoms or groups of atoms (called **functional groups**) that important molecules are formed with specific properties that behave in characteristic ways. The carbon skeleton forms the backbone and the functional group defines the type of compound that is formed. The compounds formed can be divided into **families** with similar properties.

When a functional group is attached to a carbon backbone the properties of the new molecule formed are often due to the difference in electronegativity between the atoms in the functional group and the carbon atoms. The dipole set up in the new molecules affects the melting and boiling point of the compound, its solubility properties and its reactivity.

Alcohols: functional group: hydroxyl -OH

One of the most important functional groups is the hydroxyl (-OH) group which is found in the family of molecules known as alcohols. An alcohol molecule can be considered as a water molecule which has had one of the hydrogen atoms replaced by a carbon chain (or R group).

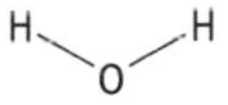

H O H
water

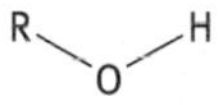

R O H
alcohol

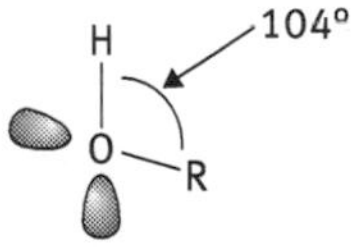

H
104°
O
R

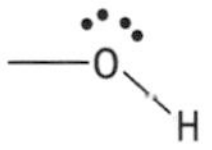

—O H
hydroxyl group

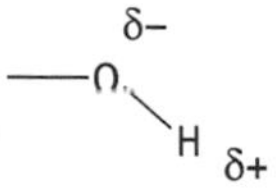

δ−
—O H
δ+

sp³ hybridised oxygen atom with two lone pairs of electrons

As in the water molecule, the oxygen atom has two lone pairs of electrons and is electronegative, making the hydrogen electropositive. It is this property which makes alcohols polar and they undergo hydrogen bonding in a similar way to water molecules. Alcohols can hydrogen bond with molecules of the same type and with water.

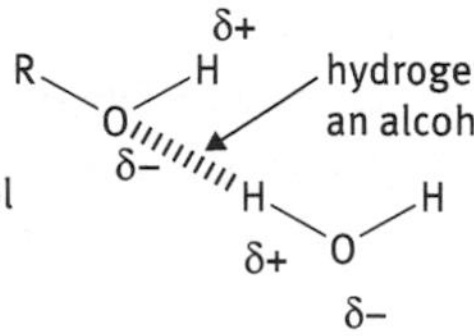

δ+
R O H
δ−
alcohol
H O H
δ+
δ−
water

hydrogen bond formation between an alcohol molecule and water

δ+
R—O—H   hydrogen bond formation between alcohol molecules
δ−
alcohol
H—O—R
δ+   δ−
alcohol

The physical properties of alcohols are as a result of hydrogen bonding. This accounts for their relatively high boiling points and the fact that they are good solvents.

Common alcohols and their boiling points are:

methanol CH_2OH	65°C
ethanol CH_3CH_2OH	78°C
propanol $CH_3CH_2CH_2OH$	97°C
butan-1-ol $CH_3CH_2CH_2CH_2OH$	117°C

When the –OH group is attached to one of the sp^2 hybridised carbon atoms of the benzene ring, a family of compounds known as phenols is formed. The simplest member of this family is phenol itself, C_6H_5OH

phenol

Thiols: functional group: sulphydryl –SH

Thiols are the sulphur analogues of alcohols. They are characterised by the sulphydryl (SH) group. Thiols themselves have unpleasant smells and ethanethiol is added to natural gas to produce the typical smell which alerts us to the presence of the gas.

Thiols have lower boiling points than the equivalent alcohols because the sulphur atom has a lower electronegativity than oxygen, and so the molecules are less polar. Thiols cannot undergo hydrogen bonding to the same extent as alcohols for the same reason. The –SH group is found in the α-amino acid cysteine in which it performs the important function of linking two peptides through a disulphide (-S-S-) bridge and so enabling three-dimensional structures to be formed.

Carbonyls: functional group: C=O

The carbonyl (C=O) group is part of the structural formula of a number of different organic families. The carbonyl group consists of an sp^2 hybridised carbon atom attached via a double bond to an oxygen atom.

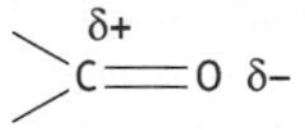

$$>C^{\delta+}=O^{\delta-}$$

carbonyl group

The geometry at the carbon atom is therefore planar and the electronegative oxygen atom withdraws electron density from the carbon atom and induces a strong dipole in the C=O bond. The carbon atom is left with a slight positive charge which makes it very reactive.

Aldehydes and ketones

If one of the atoms bonded to the carbonyl carbon atom is a hydrogen atom (and the other a carbon) we have the family of molecules known as **aldehydes**.

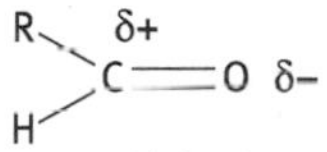

$$(R)(H)C^{\delta+}=O^{\delta-}$$

aldehyde

If the carbonyl carbon atom is attached to two other carbon atoms the compound is a **ketone.**

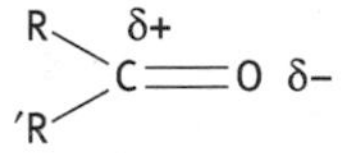

$$(R)('R)C^{\delta+}=O^{\delta-}$$

ketone

Some of the simpler aldehydes and ketones are given below, together with their boiling points.

Aldehydes		
HCHO	methanal (formaldehyde)	-21°C
CH_3CHO	ethanal	20°C
CH_3CH_2CHO	propanal	49°C

Ketones		
CH_3COCH_3	propanone (acetone)	56°C
$CH_3COCH_2CH_3$	butanone	80°C
$CH_3COCH_2CH_2CH_3$	pentan-2-one	102°C
$CH_3CH_2COCH_2CH_3$	pentan-3-one	102°C

The large dipole moment in the carbonyl group results in relatively high intermolecular forces and therefore boiling points, but the molecules cannot undergo hydrogen bonding with each other. They can, however, undergo hydrogen bonding with water which makes the smaller compounds soluble in water and also good solvents for many polar and non-polar molecules.

Carboxylic acids: functional group: carboxyl group –COOH

Carboxylic acids contain the carboxyl group which consists of an sp^2 hybridised carbon atom connected to an oxygen atom by a double bond (i.e. a carbonyl group) and a hydroxyl (-OH) group. Again the geometry at the carbon atom is planar.

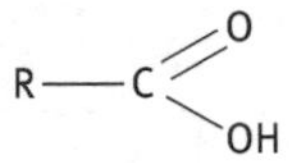

R—C(=O)OH carboxyl group

The family of molecules containing the carboxyl group are known as carboxylic acids and are weak acids. This means that they show acidic behaviour and can dissociate to lose a hydrogen ion from the hydroxyl group. This occurs due to hydrogen bonding between the hydroxyl hydrogen and a neighbouring water molecule, as shown below. A very small percentage of these interactions result in the hydrogen atom being pulled from the carboxyl group to leave a **carboxylate** anion.

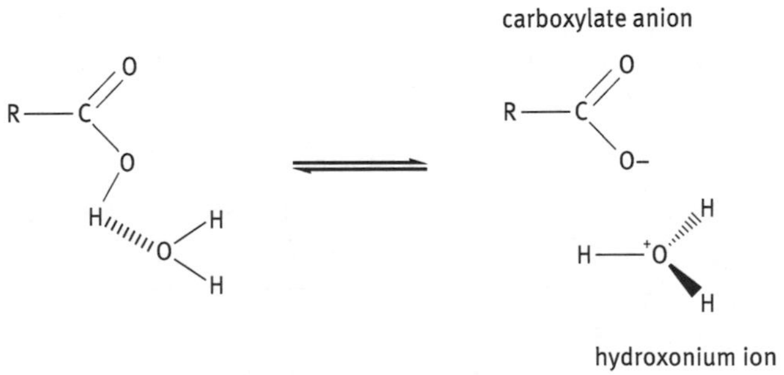

carboxylate anion

hydroxonium ion

$$R\text{-}COOH + H_2O \rightleftharpoons R\text{-}COO^- + H_3O^+$$

Many simple carboxylic acids are well-known compounds present in familiar substances:

Formula	Name	Common name	Found in
HCOOH	methanoic acid	formic acid	ant stings
CH_3COOH	ethanoic acid	acetic acid	vinegar
CH_3CH_2COOH	propanoic acid	propionic acid	Swiss cheese and sweat
$CH_3CH_2CH_2COOH$	butanoic acid	butyric acid	rancid butter

Several other families of organic molecules can be derived from carboxylic acids. Some of the most important biological ones are described below.

Esters: functional group: -COO-R′

Esters are formed from carboxylic acids and alcohols by esterification, or replacement of the hydroxyl hydrogen atom of the carboxyl group by the carbon-containing chain (alkyl group) of the alcohol group; the –C-O-C- bonds are known as ester linkages. Most esters have pleasant odours and flavours and are responsible for the tastes and fragrances of fruits and drinks.

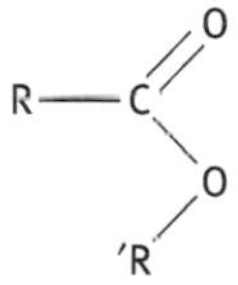

O
R—C
O
′R

Amides: functional group: amide group -CONH$_2$

The amide group consists of a carbonyl group directly attached to a nitrogen atom.

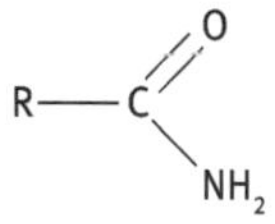

O
R—C
NH_2

Because of the strong electronegativity of both the oxygen atom on the carbonyl group and the nitrogen atom of the amino group, amides are highly polar and form very strong intermolecular interactions. Their melting and boiling points are higher even than those of carboxylic acids. They are also highly water soluble because of the dipole–dipole and hydrogen bonding interactions which occur with water.

TAKING IT FURTHER:
The peptide bond
(p. 28)

The amide bond between a carbonyl carbon atom and an amino group is known as a peptide bond when it links amino acids together into polypeptides.

Amines: functional group: amino group -NH_2

Amines are molecules which contain the amino group. The amino group is derived from ammonia, NH_3. It comprises an sp^3 hybridised nitrogen atom linked, in the case of primary amines, to two hydrogen atoms. The nitrogen atom possesses a lone pair of electrons and this makes it electronegative and sets up dipoles in the N-H bonds. Amines can therefore undergo hydrogen bonding with other molecules of amine and with water. This makes them soluble in water. Solutions of amines in water are basic because a small percentage of the amine molecules can remove a hydrogen atom from a water molecule leaving a hydroxide (OH^-) ion. Don't confuse this with the hydroxyl (-OH) group which is not charged.

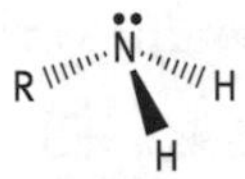

amine

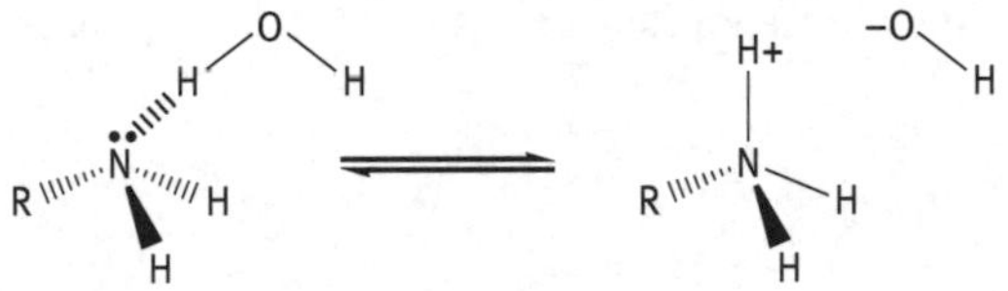

Amino groups are found in amino acid chains along with carboxyl groups.

Many molecules are known in which some, or all, of the hydrogen atoms of the amino group are replaced by carbon chains or rings. This can lead to –NH and –N groups. These groups are very common and very important biologically and are found in a variety of different types of molecules. They are also found in many molecules of pharmacological importance such as alkaloids (morphine, codeine, heroin), stimulants (caffeine and nicotine), amphetamines and barbiturates.

Phosphoric acid and phosphates: functional group: HPO_4^{2-}

Phosphoric acid is a relatively strong acid which dissociates in a stepwise fashion to give the HPO_4^{2-} ion:

phosphoric acid H_3PO_4 ⇌ hydrogen phosphate ion HPO_4^{2-} + $2H^+$

The HPO_4^{2-} ion can link carbon atoms of sugars to give phosphoester bonds. This linkage is analogous to the equivalent all-carbon ester bonds. In this case there are two phosphorus to carbon ester linkages so this is known as a **phosphodiester** bond.

phosphoric group

phosphodiester group

5.8 Summing up

1. Prior to forming covalent bonds, carbon undergoes hybridisation, in which 2s atomic orbital electrons are promoted to the same energy level as the 2p atomic orbital electrons.
2. sp^3 hybridisation involves the formation of four hybrid atomic orbitals, each containing one unpaired electron; carbon is therefore capable of forming four covalent bonds.
3. Carbon readily forms covalent bonds with itself, leading to a variety of possible structures, including ring and chain structures.
4. The tetravalency of carbon is important in determining the shapes of biological molecules.

5.9 Test yourself

The answers are given on p. 180.

Question 5.1
Explain the difference between a carbon atom which is sp^3 hybridised and one which is sp^2 hybridised.

Question 5.2
How would you explain the three-dimensional shape of the methane (CH_4) molecule?

Question 5.3
Which of the following hydrocarbons has a carbon–carbon double bond in its structure?
(a) C_3H_8
(b) C_2H_6
(c) C_2H_4
(d) CH_4

Question 5.4
In which of these ring structures would you find sp^2 hybridised carbon atoms?

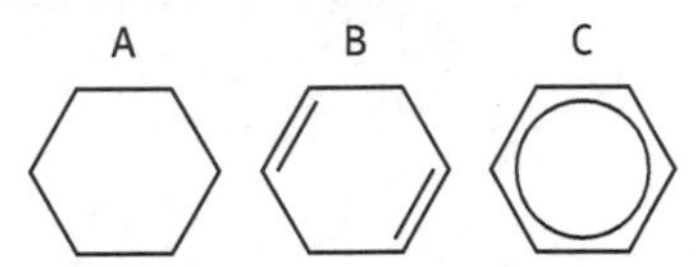

Taking it further

Carbon structures

The tetravalency of carbon and its different hybridisation states make a variety of shapes of molecules possible. Carbon readily bonds to other elements but also has a propensity to bond to itself, forming ring structures and chain structures. There are three known pure carbon structures, each of which has some remarkable properties. **Graphite** and **diamond** occur naturally; **fullerene** structures were only discovered in 1985. The different forms of the same element are called **allotropes**.

Structure of graphite

Naturally occurring graphite is called beta-graphite and it comes in a hexagonal form (Fig. 45). The hexagonal form of graphite has carbon atoms arranged in a hexagon; the hexagons form a plane. Each carbon atom in the hexagon is attached to three other carbon atoms. Carbon in graphite shows sp^2 hybridisation. The unhybridised p orbitals all lie parallel to each other and perpendicular to the graphite plane, generating a sea of delocalised pi electrons across this planar ring structure. The sheet-like structure with weak interlayer forces makes graphite a very soft substance; the layers can slide easily across each other and graphite is used as a lubricant.

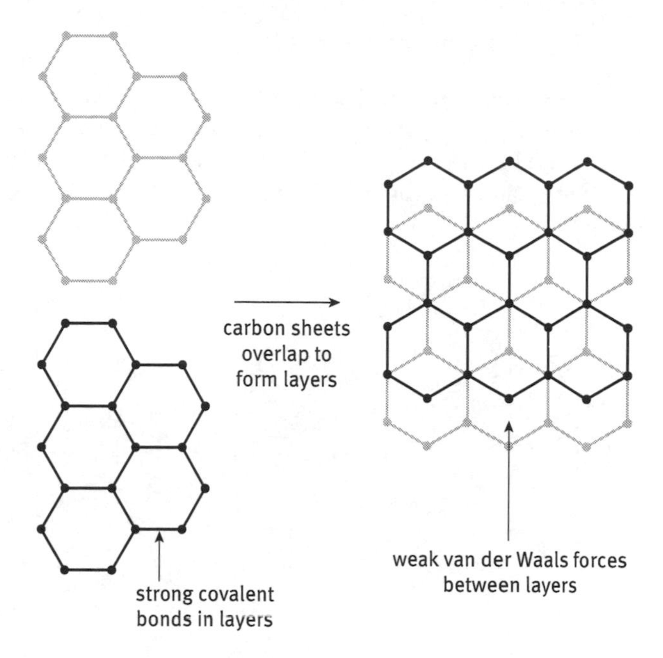

Figure 45. Hexagonal graphite structure

Structure of diamond

In diamond, each carbon atom adopts a sp^3 hybridisation; each carbon in the diamond structure is therefore at the centre of a tetrahedron, and diamond itself is a crystalline lattice (Fig. 46).

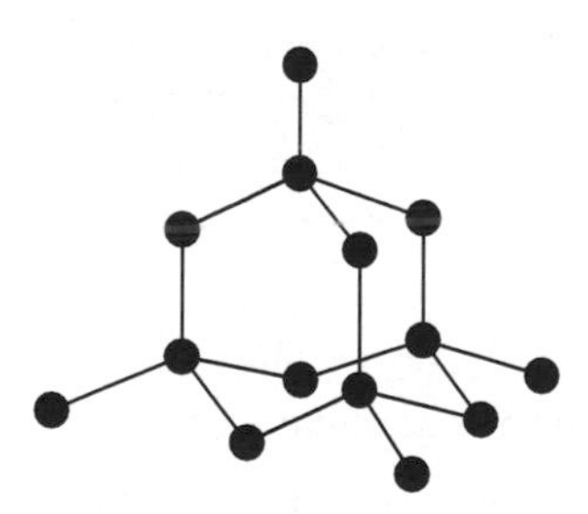

Figure 46. Diamond structure

Each carbon atom is covalently bonded to four other carbon atoms. The sigma bonds formed have maximum electronic overlap with each other. The structure formed is therefore very rigid. This makes diamond one of the hardest naturally occurring substances.

Structure of fullerenes

The simplest type of fullerene is made up of 60 carbon atoms arranged in a series of interlocking hexagons and pentagons, forming a structure that looks like a soccer ball, and often referred to as 'buckyballs' by the 1996 Nobel Laureates Curl, Kroto and Smalley, who first discovered them (Fig. 47).

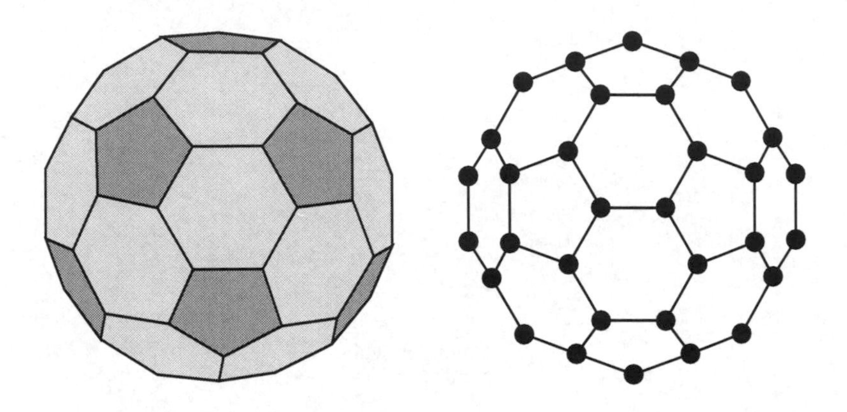

Figure 47. Simple fullerene structures

These are the only known molecules, composed of a single element, to form a hollow spheroid structure.

The pentagonal ring structures in the fullerenes, which are absent in graphite and diamond, provide the means for generating these curved structures.

Fullerene structures are providing some exciting ideas in biology. The hollow nature of this structure offers the potential for filling it, and perhaps using it as a novel drug-delivery system. Drugs could also be attached to the surface of the 'buckyball'; the spheroid structure might be expected to gain ease of entry to active sites of enzymes!

A 60-carbon fullerene is about one nanometre (10^{-9} m) in diameter, roughly the size of many small pharmaceutical molecules (in comparison, a human hair is as wide as 50 000 buckyballs). The fullerenes' unique structures might also be used as a scaffolding for building drug molecules.

A significant spin-off product of fullerene research are **nanotubes**, based on carbon or other elements (Fig. 48). These systems consist of graphitic layers seamlessly wrapped to form cylinders. They are only a few nanometres in diameter, but up to a millimetre long. Among the highlights of nanotube research to date is the demonstration that tubes can be opened and filled with a variety of materials, including biological molecules.

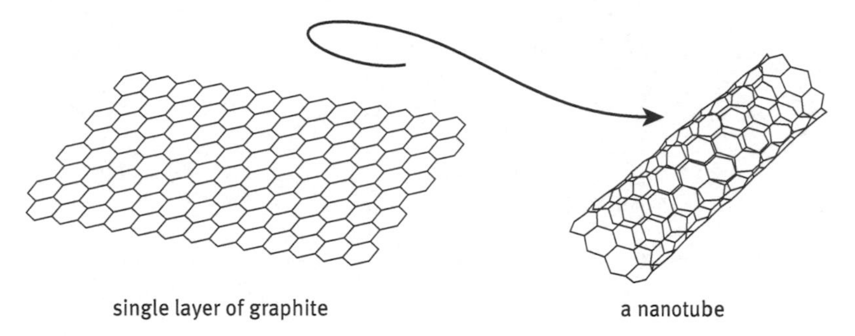

This figure is reproduced with permission of the copyright holders: Professor Charles M. Lieber Research Group

Figure 48. Carbon nanotubes.

Nanoscience is a relatively new area of science, promising exciting new developments. Nanoscience is developing new power devices, sensors, new means of data storage, molecular electronics and nanotube gears to drive nanomachines! Just a taste of things to come!

Taking it further

Macromolecules

In biology, macromolecules (biopolymers) include carbohydrates, proteins and nucleic acids; lipids are often included although they are not a polymer. Macromolecules often define cells, conveying enzymatic activity, genetic inheritance, cell growth, energy storage and conversion, and interaction with the environment. Together they illustrate the commonality of life on earth; the way they are used in different forms and combinations explains today's biodiversity.

Polymers are made up of repeating units, monomers, that may be identical or different (but of a similar chemical type); simple sugar monomers (e.g. glucose, galactose) combine to form complex polysaccharides (e.g. glycogen, starch), amino acid monomers combine to form polypeptides (that fold to produce proteins), and nucleotide monomers combine to form nucleic acids.

Monomers are linked together in a condensation reaction, a dehydration synthesis reaction, with the removal of a water molecule; in many cases the polymer can be taken apart by a 'reverse' hydrolysis reaction, the addition of water. Specifically there must be a hydroxyl group on one molecule, and a hydrogen on another molecule, that is typically bonded to an electronegative atom (e.g. O or N); removing OH and H produces water (dehydration), while the pair of electrons left are responsible for forming the bond (synthesis) between the two reactants. For example:

$$-\mathrm{C}-\mathrm{OH} + \mathrm{HO}-\mathrm{C}- \text{ react to give } -\mathrm{C}-\mathrm{O}-\mathrm{C}- + \mathrm{H}-\mathrm{O}-\mathrm{H}$$

$$-\mathrm{C}-\mathrm{OH} + \mathrm{HN}-\mathrm{C}- \text{ will react to give } -\mathrm{C}-\mathrm{N}-\mathrm{C}- + \mathrm{H}-\mathrm{O}-\mathrm{H}$$

Dehydration synthesis can occur between a carboxyl group and a hydroxyl group:

$$\mathrm{O}=\mathrm{C}-\mathrm{OH} + \mathrm{HO}-\mathrm{C}- \text{ to give } \mathrm{O}=\mathrm{C}-\mathrm{O}-\mathrm{C}- + \mathrm{H}-\mathrm{O}-\mathrm{H}$$

$$-\mathrm{C}(=\mathrm{O})\mathrm{OH} + \mathrm{HO}-\mathrm{C}\lessdot \longrightarrow -\mathrm{C}(=\mathrm{O})-\mathrm{O}-\mathrm{C}\lessdot + \mathrm{H}_2\mathrm{O}$$

This is how fatty acids (the carboxyl group) are bound to glycerol (which supplies the hydroxyl group) to form triacylglycerols (lipids).

Or,

$$\mathrm{O}=\mathrm{C}-\mathrm{OH} + \mathrm{H_2N}-\mathrm{C}- \text{ gives } \mathrm{O}=\mathrm{C}-\mathrm{N(H)}-\mathrm{C}- + \mathrm{H}-\mathrm{O}-\mathrm{H}$$

$$-\mathrm{C}(=\mathrm{O})\mathrm{OH} + \mathrm{H_2N}-\mathrm{C}\lessdot \longrightarrow -\mathrm{C}(=\mathrm{O})-\mathrm{N(H)}-\mathrm{C}\lessdot + \mathrm{H}_2\mathrm{O}$$

TAKING IT FURTHER: The peptide bond (p. 28)

This results in the formation of a peptide bond that links amino acids together.

Carbohydrate monomers are linked together to generate a polymer, a polysaccharide chain (Fig. 49). The macromolecule may grow by adding monomers to the end of the chain, or by branching off the chain. The -C-O-C- covalent linkage is referred to as a **glycosidic** bond; the 'chain links' are called 1,4-glycosidic bonds (A), since they involve the C1 and C4 of two monomers, while the 'branch links' are 1,6-glycosidic bonds (B), involving C1 and C6 of two monomers. Figure 49 is composed solely of glucose monomers; the structure derived from this would be glycogen, a secondary long-term energy store in animals. At the anomeric carbon of the glucose ring (see Fig. 57), the hydroxyl group may be shown pointing up above the plane of the ring, in which case the glycosidic bond would be designated 'β', whereas in Fig. 49 the glycosidic bonds shown would be designated 'α'.

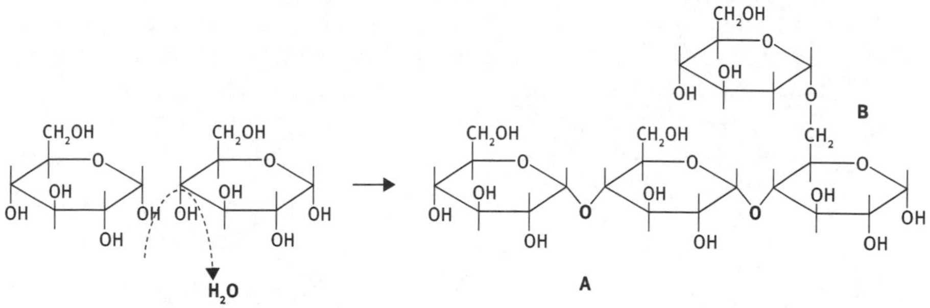

Figure 49. Growing a polysaccharide chain

Fats (lipids) are an important energy store in animals; the triacylglycerol components are formed from the addition of individual fatty acids to a glycerol backbone, three fatty acids in total, hence ‘tri’ in triacylglycerol (Fig. 50).

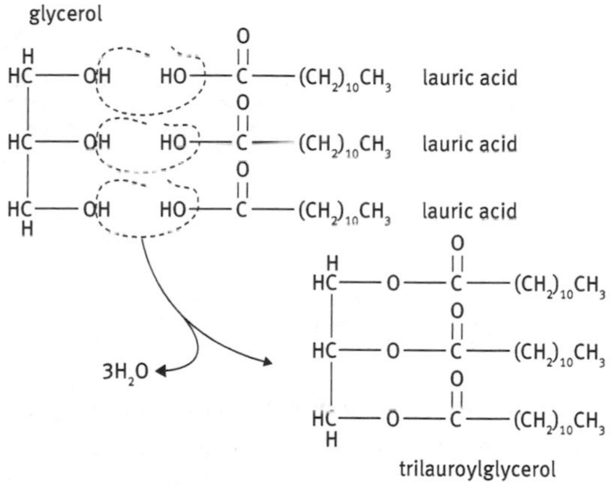

Figure 50. Forming trilauroylglycerol

A fatty acid, which may be the same or different, is added to each hydroxyl group on the glycerol, through a condensation reaction in which water is formed. The triacylglycerol is not a polymer as such; however, given the rather hydrophobic nature of the molecule (by virtue of the long hydrocarbon fatty acid ‘tails’), these molecules associate to form the fat deposits in animal tissues. In this case, the -C-O-C- bonds formed are referred to as **ester** bonds. Esters are formed by condensing an acid with an alcohol, and most commonly from carboxylic acids and alcohol. Esters are ubiquitous; naturally occurring lipids are the fatty acid esters of glycerol, as shown in Fig. 50.

Phosphoesters form the backbone of nucleic acid molecules. Figure 51 shows how starting with three ingredients, phosphate, sugar (deoxyribose) and one of the four DNA bases, thymine, we can build the DNA backbone, comprising a repeating sugar–phosphate polymer; bases attached to the sugar ultimately cross-link with bases on an opposing strand to produce the double-stranded DNA helix.

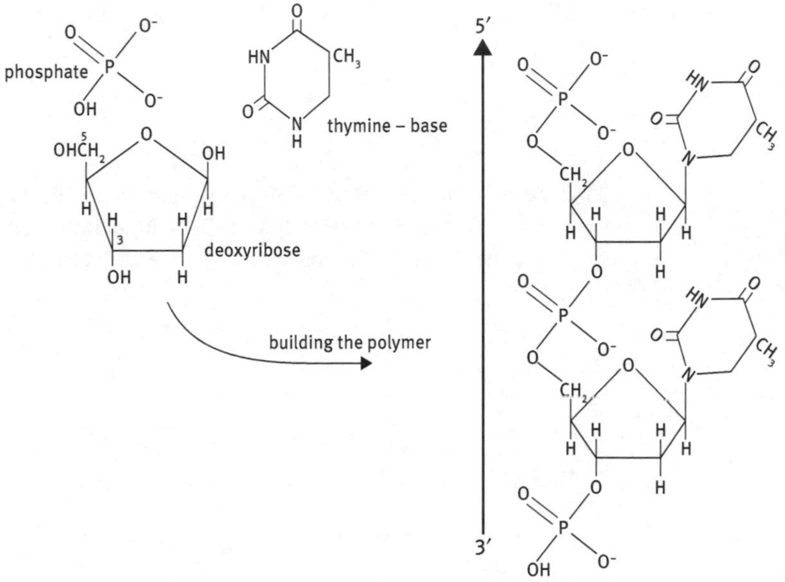

Figure 51. Growing a DNA molecule

Phosphate forms an ester bond through the hydroxyl groups at C3 and C5 of the deoxyribose sugar; in effect it forms a phospho**di**ester bond, linking two sugars together. The phosphate linking of the deoxyribose at the 3 and 5 carbon positions gives the polymer a 'direction', the 3-prime and 5-prime ends. This directionality of the molecule is biologically vital in terms of its replication and reading (translation) of the genetic code.

In Section 2.11, we looked at the characteristics of the peptide bond, also formed in a condensation reaction, and responsible for linking amino acids together into a polypeptide polymer. Through such polymer formations, biology produces the macromolecular hallmarks of life.

06 The same molecule but a different shape

BASIC CONCEPTS:

Molecules with the same chemical formula, but different structure and characteristics, are isomers. The tetravalency of carbon predicts the existence of mirror image isomers, so-called stereoisomers. Organisms in general impose a 'chiral environment'; for example we only use amino acids in their L-form, or sugars in their D-form. Enzymes will usually only recognise one type of isomer. The introduction of a 'wrong' isomer into the body can have disastrous consequences.

6.1 Isomers

In chemistry we define isomers as ' two or more different compounds with the same chemical formula but different structure and characteristics'.

Isomers can take many different forms, but there are two main forms of isomerism, namely **structural isomerism** and **stereoisomerism**.

In **structural isomers**, the atoms and functional groups are joined together in different ways. For example, the molecular formula C_2H_6O is that of ethanol, but it is also that of dimethylether (CH_3OCH_3). Both have the same elemental composition but are not the same compound; they are structural isomers.

In **stereoisomers** the bond structure is the same, but the geometrical positioning of atoms and functional groups in space differs. This class includes **optical isomerism** where different isomers are mirror images of each other, and **geometric isomerism** where functional groups at the end of a chain can be twisted in different ways.

6.2 Optical isomerism

As we have previously seen, carbon is a tetravalent atom capable of forming four covalent bonds. Once formed these bonds point towards the centre of a tetrahedron.

The tetrahedral carbon predicted the existence of **mirror image isomers.** When carbon makes four single bonds with four different groups, non-superimposable mirror image molecules (**enantiomers**) exist (Fig. 52).

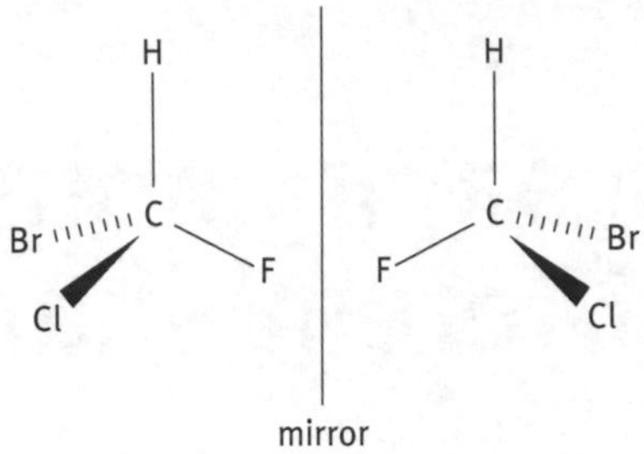

Figure 52. Non-superimposable mirror image structures

Derived from the Greek word *enantio* meaning opposite, enantiomers are non-superimposable mirror image structures.

Enantiomers are particularly important in biology. A molecule containing an **asymmetric**, or **chiral**, carbon can adopt two structures. An asymmetric carbon has four different atoms or groups attached to it. The amino acid structures shown in Fig. 53 are mirror images of each other. They have essentially identical chemical and physical properties. Physically they can be distinguished by their ability to rotate plane polarised light by equal amounts but in opposite directions (hence the D or L nomenclature). A solution of a D-isomer will rotate plane polarised light to the right (dextrorotatory = right-handed = clockwise); a solution of the L-isomer will rotate plane polarised light to the left (levorotatory = left-handed = counterclockwise). A solution of equal amounts of D- and L-isomers will not rotate plane polarised light; such a mixture is called a **racemic** mixture.

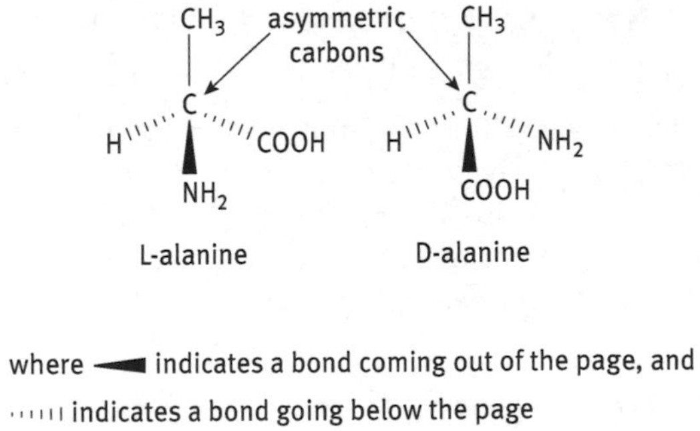

Figure 53. D- and L-alanine are enantiomers

Sugars can also exist in D- and L-forms. The simplest three-carbon sugar is glyceraldehyde, shown in Fig. 54 in its D- and L-forms.

There is only one asymmetric (chiral) carbon in this structure. Looking at this molecule with the aldehyde group (-CHO) at the top and 'furthest away', if the OH group to the asymmetric carbon is on the right, then it is the D-form. If it is on the left it is the L-form.

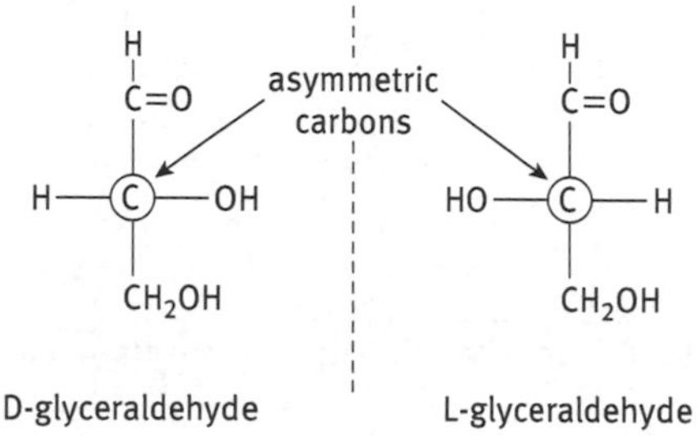

asymmetric carbons

D-glyceraldehyde L-glyceraldehyde

Figure 54. D- and L-glyceraldehyde are enantiomers

In more complex sugars the number of chiral carbons increases and consequently the number of possible isomers also increases. There are in fact 2^n stereoisomers possible for each chiral centre (n) in the molecule.

D-glucose and L-glucose are enantiomers, they are mirror images of each other. In glucose there are four chiral centres (carbons 2, 3, 4 and 5 in Fig. 55). In determining whether the structure is the D- or L-form, only the chiral carbon 'furthest' from the aldehyde group is considered (carbon 5 in Fig. 55). In D-glucose, the position of each group at each of the chiral carbons is 'reversed' relative to L-glucose; D- and L-glucose are therefore mirror images of each other.

Enantiomers
stereoisomers that are mirror images of each other

D-glucose L-glucose

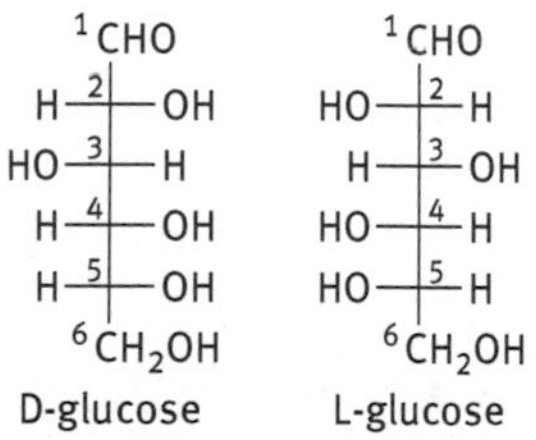

Figure 55. D- and L-glucose are enantiomers

On the other hand, D-glucose and D-galactose are referred to as **diastereomers**; they are stereoisomers that are not mirror images of each other. In Fig. 56, the determination of D- or L-forms is still made by considering the position of the OH group at the chiral carbon furthest from carbon atom 1, thus both the glucose and galactose structures shown are D-forms (the OH group is on the right at carbon atom 5). However, D-glucose and D-galactose are clearly not mirror images of each other.

Diastereomers
stereoisomers that are not mirror images of each other

D-glucose	D-galactose
^{1}CHO	^{1}CHO
H–2–OH	H–2–OH
HO–3–H	HO–3–H
H–4–OH	HO–4–H
H–5–OH	H–5–OH
$^{6}CH_2OH$	$^{6}CH_2OH$

Figure 56. D-glucose and D-galactose are diastereomers

Considering the ring structure of glucose, another type of stereoisomer becomes apparent, namely **anomers.** Here the difference resides in the configuration of groups around one carbon atom, in this case carbon atom 1. In solution, glucose undergoes a process known as **mutarotation.** The open 'chain' form of glucose is in equilibrium with the ring structure, through the formation of a hemiacetal bond (Fig. 57).

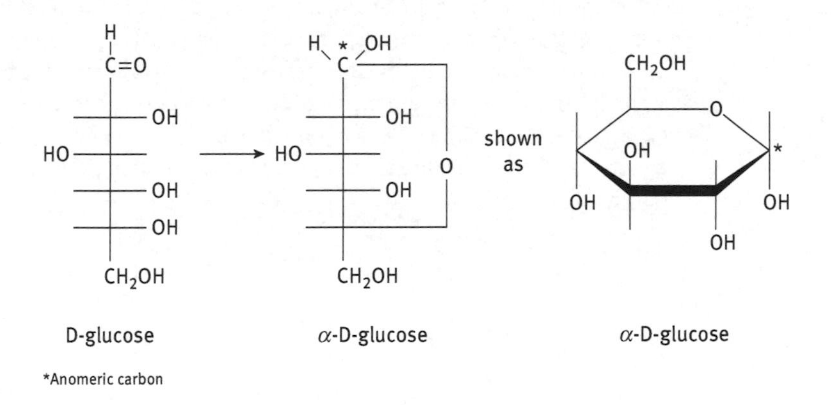

Figure 57. Mutarotation in glucose

In Fig. 57, D-glucose cyclises to form α-D-glucose, in which the hydroxyl group at the anomeric carbon (shown by *) is shown pointing down. If the hydroxyl group were shown pointing up, then this would be β-D-glucose. Thus the ring form of glucose has one additional chiral centre than does the open chain form.

Furthermore, the alpha and beta ring forms of glucose will equilibrate in solution through the open chain form (Fig. 58).

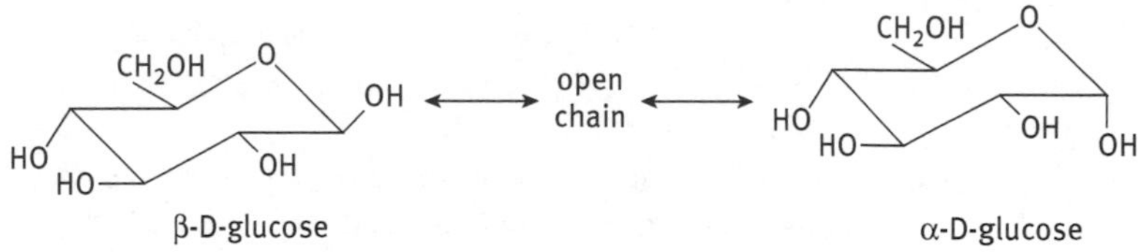

β-D-glucose ⟷ open chain ⟷ α-D-glucose

Figure 58. Equilibration of glucose anomers

In solution and at equilibrium, the beta ring form predominates (the hydroxyl groups are spaced further apart in the beta form, thus avoiding steric crowding).

6.3 Geometric isomerism

There are different types of geometric isomerism but a common one is that referred to as ***cis-trans* isomerism.** If a pair of stereoisomers contains a carbon–carbon double bond then it is possible to get a *cis* or *trans* arrangement of the substituents at each end of the double bond. These are referred to as ***cis-trans* isomers**. The simple example given here is that of butene.

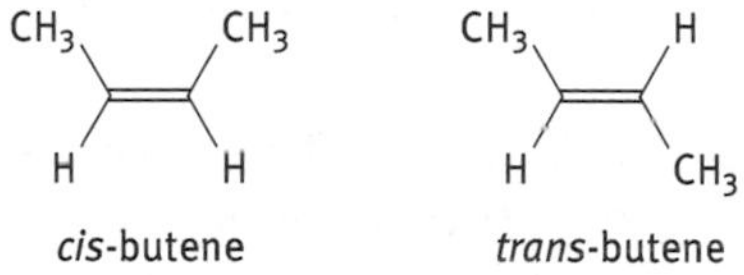

cis-butene *trans*-butene

With the methyl (CH_3) groups on the same side of the double bond, the *cis*-isomer is realised. The *trans*-isomer is realised when the methyl groups are on opposite sides of the double bond.

The carbon–carbon double bond effectively fixes the atoms or groups attached to it and prevents their rotation about the bond (remember, there is no rotation about a double bond).

In unsaturated fatty acids (i.e. those which contain a carbon–carbon double bond), orientation of groups about this bond may be either *cis* or *trans*.

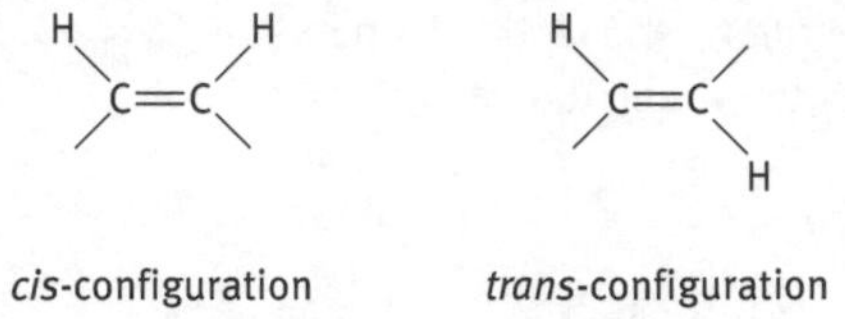

cis-configuration *trans*-configuration

A *cis* configuration causes a 'kink' or bend in the carbon chain, whereas a *trans* configuration forms a straight chain (Fig. 59).

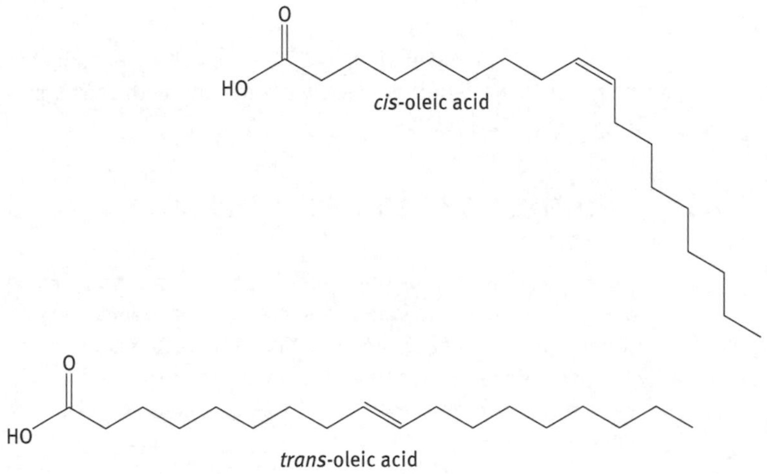

cis-oleic acid

trans-oleic acid

Figure 59. *cis* and *trans* geometric isomers of oleic acid

Naturally occurring unsaturated vegetable oils have almost all *cis* bonds, but using oil for frying causes some of the *cis* bonds to convert to *trans* bonds. If oil is constantly reused, more and more of the *cis* bonds are changed to *trans* until significant numbers of fatty acids with *trans* bonds build up. This is a health concern, since fatty acids with *trans* bonds have been shown to raise total blood cholesterol levels, thus increasing the risk of heart disease; they have also been shown to be carcinogenic, or cancer-causing.

There are, however, some striking examples in biology where the natural conversion between isomers is of central importance. In the mammalian eye, 11-*cis*-retinal (a form of vitamin A) forms part of the photoreceptor apparatus ('11' refers to the eleventh carbon atom in the chain). Exposure to light isomerises 11-*cis*-retinal to all-*trans*-retinal, in doing so triggering a series of reactions involved in the biochemical pathways of vision (Fig. 60).

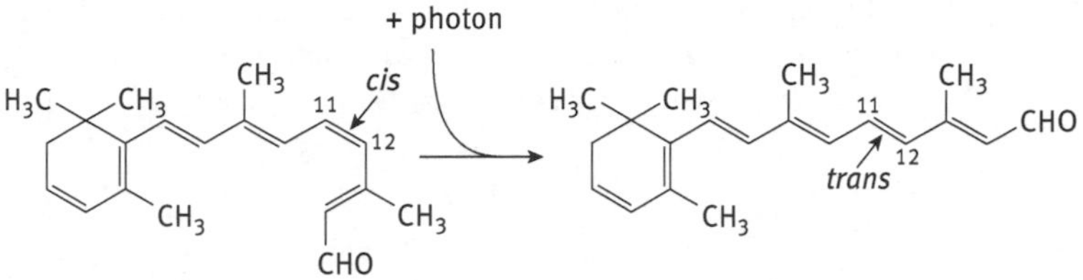

Figure 60. The structure of 11-*cis*-retinal and all-*trans*-retinal

6.4 Isomers as a problem

Life forms can distinguish isomers based on their different shapes. Normally, one isomer is biologically active and others are inactive. Our cells have imposed a **chiral environment**; in other words, only certain isomers are recognised and incorporated. We use only L-amino acids to construct our proteins, or D-sugars in carbohydrate metabolism. We favour *cis*-isomers in fatty acids and *trans*-peptide bonds in proteins. This molecular recognition resides at the level of the enzymes which orchestrate these processes. However, problems can arise when an inappropriate isomer is introduced into this environment.

In the 1960s, many pregnant women who had taken **thalidomide** gave birth to deformed babies. One of the enantiomers, isomer type 1, acted as a sedative as intended, but the other isomer, type 2, caused birth defects. Fig. 61 shows the two isomers of thalidomide which differ in the rotation of the ring group about the carbon marked *.

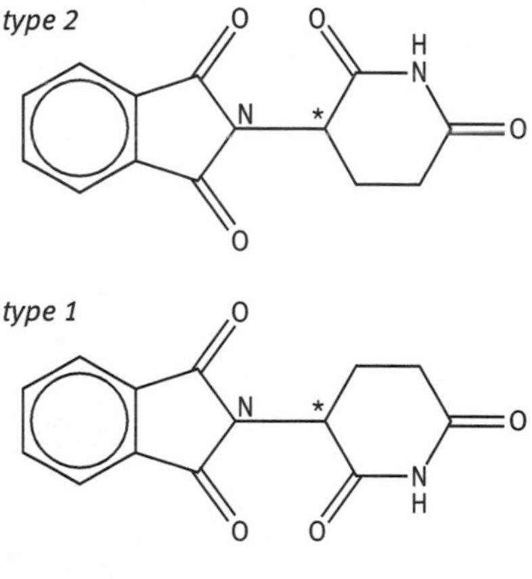

Figure 61. Two isomers of thalidomide

Sold over the counter in a number of pain remedies, **ibuprofen** is a mixture of two non-superimposable mirror image enantiomers (Fig. 62). Therapeutic activity is shown only when isomer type 2 is used.

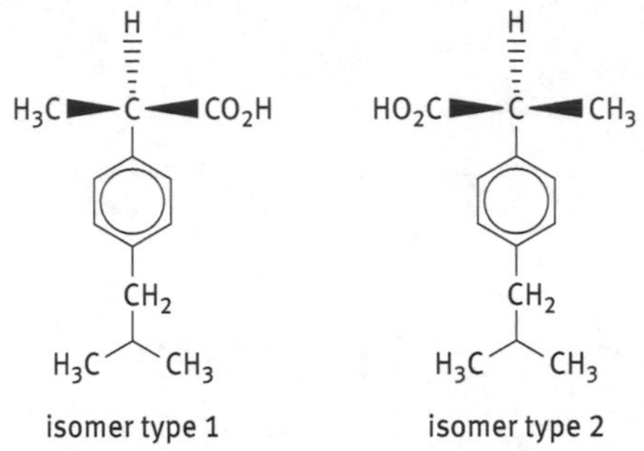

isomer type 1

isomer type 2

Figure 62. Two isomers of ibuprofen

In today's pharmaceutical industry, legislation dictates that the production of new drugs must result in just one isomeric form to be marketed. Investment in new methods of chiral synthesis and methods of isomer separation have resulted in more potent and safer drugs.

6.5 Summing up

1. Isomers are defined as 'two or more different compounds with the same chemical formula but different structure and characteristics'.
2. By virtue of the tetravalency of carbon, stereoisomers (mirror images) exist for many biological molecules, including amino acids and sugars.
3. Molecules may contain one or more asymmetric (chiral) carbon atoms; the arrangement of groups around such atoms gives rise to different isomers, including enantiomers (non-superimposable mirror images), diastereomers (which are not mirror images), and anomers (which vary in their configuration of one group about one asymmetric carbon).
4. Biological life forms impose a chiral environment, in which only certain isomers are tolerated, e.g. L-amino acids and D-sugars.

6.6 Test yourself

The answers are given on p. 180.

Question 6.1
Choose the phrase which correctly describes the relationship between the molecules A and B:
(i) they are structural isomers
(ii) they are geometric isomers
(iii) they are enantiomers
(iv) they are isotopes

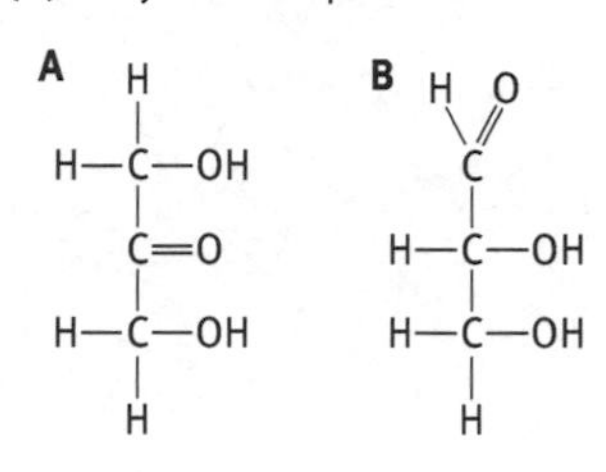

Question 6.2
What is the difference between pairs of molecules which are either enantiomers or diastereomers?

Question 6.3
What do you understand by a chiral centre?

Question 6.4
In a molecule with six chiral carbons, how many possible stereoisomers of that molecule could exist?

Question 6.5
How would you distinguish separate solutions of D- and L-glucose?

07 Water – the solvent of life

BASIC CONCEPTS:

Water is the major component of living cells, comprising around 70% of a cell's total weight. The abundance of water in biological systems inevitably dictates the behaviour of the biological molecules it interacts with. Water may behave as both an acid and a base; the autoionisation of water is fundamental to an understanding of the acid–base behaviour of biological molecules and to pH control in biological systems.

In Chapter 3 we briefly introduced some very special properties of water, namely the ability of water molecules to undergo hydrogen bonding to one another. Hydrogen bonding between water molecules is a result of the charge distribution in the water molecule, imparted by the relatively high electronegativity of the oxygen atom compared to the two hydrogen atoms. The distribution of charge and the molecule's bent geometry give rise to a **dipole moment** in the water molecule (Fig. 63).

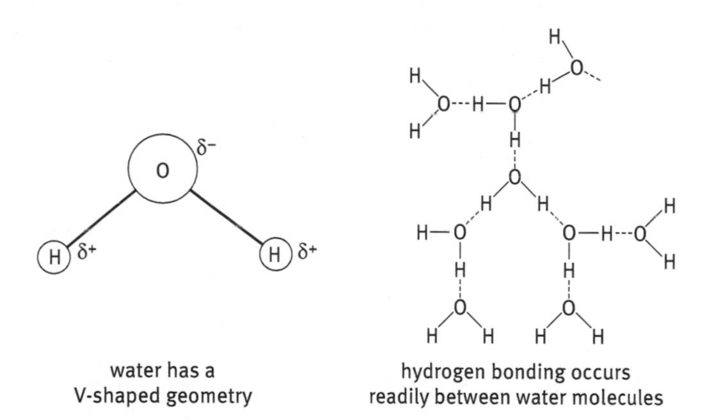

Figure 63. Dipoles and hydrogen bonding in water

7.1 Bonding in the water molecule

Oxygen has an electronic configuration of $1s^2 2s^2 2p^4$. In forming covalent bonds with hydrogen, the oxygen atom undergoes sp^3 hybridisation. Ignoring the 1s electrons, the electron configuration of oxygen can be shown using the 'electrons in boxes' notation.

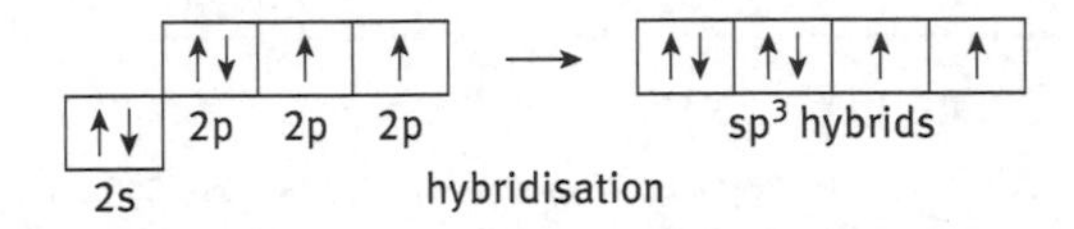

After hybridisation of the 2s and 2p orbitals, two unpaired electrons are available for bonding to hydrogen and two lone pairs of electrons remain in sp^3 hybridised orbitals on the oxygen atom (Fig. 64). The oxygen atom therefore forms two covalent bonds with hydrogen atoms by overlap of its half-filled sp^3 orbitals with the 1s orbitals of the hydrogen atoms. In fact, each water molecule is capable of forming a maximum of four hydrogen bonds to other water molecules. The two hydrogen atoms can each undergo hydrogen bonding with an oxygen atom of a neighbouring water molecule, and the oxygen atom (with two lone pairs of electrons) can hydrogen bond to hydrogen atoms on neighbouring water molecules.

REMINDER

Hydrogen bonds can form whenever a strongly electronegative atom (O, N or F) approaches a hydrogen atom in a nearby molecule that is covalently attached to a second strongly electronegative atom

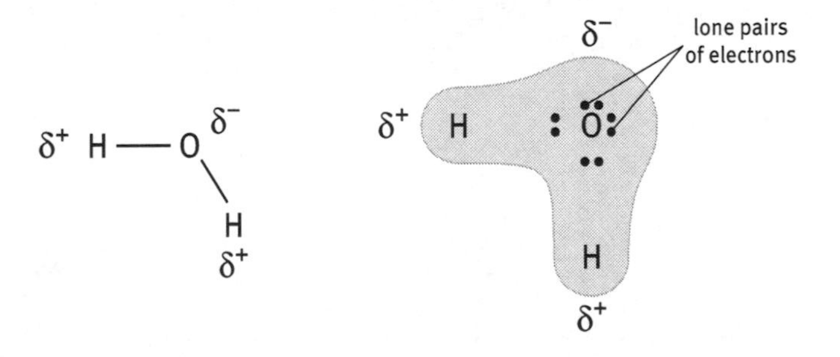

Figure 64. Electron distribution in the water molecule

7.2 The dissociation (auto-ionisation) of water

Atoms within molecules are in constant motion, and one form of motion that atoms undergo is called vibration. Vibrational motion is the stretching and shortening of a bond as the atoms move closer together and then apart. Within liquid water, at any instant in time, the hydrogen atom of a hydrogen bond may find itself closer to the oxygen atom of a neighbouring water molecule than to the oxygen atom of its own molecule (Fig. 65a). The covalent O-H bond in the H_2O molecule stretches and the hydrogen bond between the molecules shortens (Fig. 65b). Eventually the covalent O-H bond stretches so far that it breaks and a new O-H bond is formed with the neighbouring molecule (Fig. 65c). The electrons of the hydrogen atom are left behind and the lone pair of electrons, formerly in the sp^3 orbital on the oxygen atom, is used in bonding to the hydrogen atom in a dative covalent bond. This process creates two new species, each of which has a full electric charge; a **hydroxonium ion,** H_3O^+, and a **hydroxide ion,** OH^-.

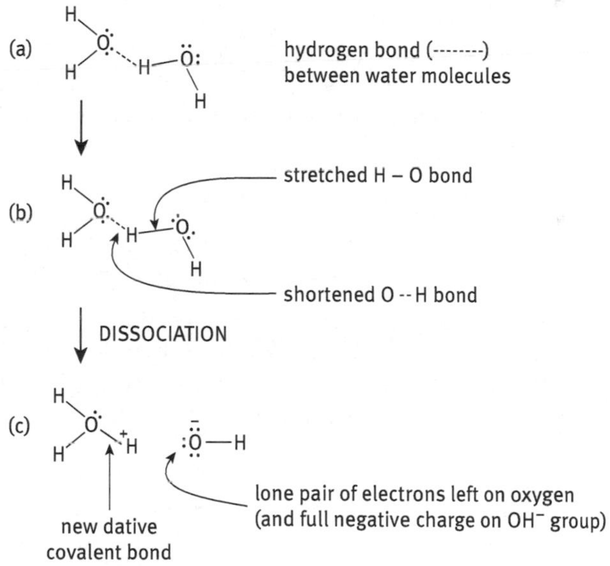

(a) hydrogen bond (--------) between water molecules

(b) stretched H – O bond

shortened O -- H bond

DISSOCIATION

(c) new dative covalent bond

lone pair of electrons left on oxygen (and full negative charge on OH^- group)

Figure 65. The dissociation of water

The process is known as the **dissociation** (or auto-ionisation) of water and scheme 1 represents the reaction (Fig. 66). The process is sometimes abbreviated, as in scheme 2. In pure water at room temperature only about one in a billion water molecules reacts in this way. The reactions in schemes 1 and 2 use the symbol $\rightleftharpoons$ to indicate that not all the molecules dissociate. The symbol $\rightleftharpoons$ is used for chemical reactions which have not gone to completion but reach an equilibrium. In reactions which reach equilibrium, only a certain number of reactant molecules (on the left-hand side of the equation) react and form products. However, when equilibrium is reached the forward and backward reactions occur at the same rate.

Scheme 1

$$H_2O + H_2O \rightleftharpoons H_3O^+ + OH^-$$

Scheme 2

$$H_2O \rightleftharpoons H^+ + OH^-$$

Figure 66. Equations which depict the auto-ionisaton of water

7.3 Acids and bases

The behaviour of water provides the basis for understanding the concept of acids and bases.

An **acid** is defined as a substance which produces hydrogen ions (H^+) by dissociation.

For example, when hydrogen chlorlde, $HCl(g)$, is added to water no HCl molecules are found, only H_3O^+ ions and Cl^- ions. Each HCl molecule reacts with water to form a hydroxonium ion and a chloride ion. The process can be represented by:

$$HCl(g) + H_2O(l) \longrightarrow H_3O^+(aq) + Cl^-(aq)$$

HCl is a strong acid because it dissociates completely in water. It is the acid which is secreted by the parietal glands in the stomach.

REMINDER

The physical states of substances are denoted by (g) = gas, (aq) = aqueous, in water solution, (l) = liquid and (s) = solid

Not all acids dissociate completely in water. Most organic acids react only partially with water to produce low concentrations of hydroxonium ions, and leave undissociated acid molecules. Such acids are called **weak acids** and the symbol $\rightleftharpoons$ is used to indicate that an equilibrium is established. Ethanoic acid (acetic acid), the acid in vinegar, behaves in this way.

$$CH_3COOH(l) + H_2O(l) \rightleftharpoons CH_3COO^-(aq) + H_3O^+(aq)$$
(ethanoic acid)

Bases are defined as substances which can extract a proton, H^+. Frequently bases extract protons from water to leave hydroxide ions, OH^-.

Bases may produce hydroxide ions directly by dissociation, for example when potassium hydroxide is added to water:

$$KOH(s) + aq \longrightarrow K^+(aq) + OH^-(aq)$$

REMINDER

Hydrogen ion concentrations are often discussed in relation to pH, despite the fact that free hydrogen ions do not exist in aqueous solution for all practical purposes. However, the hydroxonium ion will dissociate to provide a proton in situations where a proton is said to occur in theory, so the net effect is the same!

Bases which are completely dissociated in water, such as KOH, are called **strong bases.**

Other bases react with water to extract a proton and leave a hydroxide ion. For example ammonia, NH_3, is a base which reacts in this way. Because ammonia does not completely react with water and only produces a relatively small amount of hydroxide ions, ammonia is said to be a **weak base.**

$$NH_3(g) + H_2O(l) \rightleftharpoons NH_4^+(aq) + OH^-(aq)$$

REMINDER

Acids produce H^+ ions by dissociation, bases extract a proton from water to generate OH^-, or produce OH^- directly by dissociation

7.4 Using pH as a measure of acidity

The acidity of a solution is measured by the molar concentration of hydroxonium ions. These concentrations can be very small, as in bases, or very large, as in strong acids. The range of hydroxonium ion concentrations is typically from 1 M to 1×10^{-14} M. In order to convert these very small numbers to numbers which are more convenient to use, the pH scale was devised. This is a logarithmic scale in which the pH of a solution is defined by:

$$\mathbf{pH = -\log_{10}[H^+]}$$

where $[H^+]$ is the hydrogen ion concentration in mol dm^{-3} or mol l^{-1}. The unit M can also be used to indicate mol dm^{-3} or mol l^{-1}.

REMINDER

Square brackets [ ] denote 'concentration of'

$Log_{10}[H^+]$ means the logarithm to the base 10 of the hydrogen ion concentration. It is obtained on most calculators by using the 'log' function. From here on, log X will be used to indicate $\log_{10}X$.

Calculating the pH of a strong acid

0.1 M hydrochloric acid has a H^+ concentration of 0.1 M. The pH of this solution is given by: $pH = -\log[0.1]$. The value of $\log[0.1] = -1$.

Thus, pH = −(−1) = +1. The pH of this hydrochloric acid is therefore equal to 1

The pH scale can range from about 0 (strongly acidic) to +14 (very basic); a neutral solution has pH 7.0 (Fig. 67).

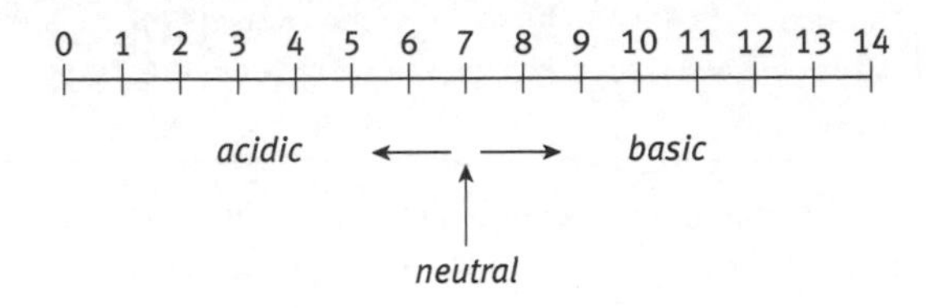

Figure 67. The pH scale

7.5 Calculating the pH of water

Water dissociates to a tiny extent. Hydrogen ions (H^+) associate with water molecules to form H_3O^+ (the hydroxonium ion).

$$H_2O + H_2O \rightleftharpoons H_3O^+ + OH^-$$

The **degree of dissociation** is measured by the equilibrium constant, which is the ratio of the concentration of the dissociated ions to the undissociated molecules. The equilibrium constant (K_{eq}) for this dissociation is defined as:

$$K_{eq} = \frac{[H^+][OH^-]}{[H_2O]}$$

Because the concentration of water, $[H_2O]$, is a constant and simply equivalent to the density of water, this equation becomes:

$$K_w = [H^+][OH^-]$$

where K_w represents the dissociation constant of water.

REMINDER

One H_2O molecule dissociates to give one H^+ ion and one OH^- ion

The concentration of hydrogen ions (hydroxonium ions) in pure water must be equal to the concentration of hydroxide ions and is found experimentally to be equal to 1.0×10^{-7} M at 25°C.

Therefore we can write $[H^+] = [OH^-] = 1.0 \times 10^{-7}$ mol

Substituting these values in the expression for K_w we find:

$$K_w = [H^+][OH^-] = 1.0 \times 10^{-7}\text{M} \times 1.0 \times 10^{-7}\text{M} = 1.0 \times 10^{-14}\ \text{M}^2$$

This tells us that very little of pure water exists in the form of its ions. It also tells us that because the value for K_w is a constant at a specific temperature, when the concentration of H_3O^+ decreases, the concentration of OH^- must increase by a similar amount, and *vice versa*.

Knowing the concentration of H_3O^+ in pure water we can calculate its pH from:

$$pH = -\log [H^+]$$

If $[H^+]$ is equal to 1.0×10^{-7} M

$$pH = -\log(1.0 \times 10^{-7} \text{ M}) = -(-7) = 7$$

Thus, the pH of pure water is 7.

7.6 The dissociation of weak acids and weak bases in water

Weak acids are substances which dissociate only partially with water to give hydroxonium ions in solution. A weak base reacts only partially with H_2O to give hydroxide ions in solution.

As for water, the extent of dissociation of weak acids and bases can be expressed by a **dissociation constant.** The dissociation constant is the equilibrium constant for the decomposition of a species into its components.

The **acid dissociation constant, K_a,** is the equilibrium constant for the reaction in which a weak acid is in equilibrium with its conjugate base and the hydroxonium ion in aqueous solution. Similarly the **base dissociation constant, K_b,** is the equilibrium constant which represents the dissociation of a base into its conjugate acid and the hydroxide ion in aqueous solution. (Conjugate means 'joined together' especially in pairs. In chemistry, a conjugate species is related to its acid or base by the difference of a proton.)

The example below shows the dissociation of ethanoic (acetic) acid. Notice that in the equilibrium expression the concentration of water is not included. This is because water is vastly in excess and the value of $[H_2O]$ is a constant.

$$\underset{\text{(weak acid)}}{CH_3COOH(aq)} + H_2O(l) \rightleftharpoons \underset{\text{(conjugate base)}}{CH_3COO^-(aq)} + H_3O^+(aq)$$

The equilibrium constant for this dissociation is equivalent to the acid dissociation constant for ethanoic acid, K_a and can be expressed by:

$$K_a = \frac{[CH_3COO^-_{(aq)}][H_3O^+_{(aq)}]}{[CH_3COOH_{(aq)}]}$$

For ethanoic acid, K_a is found to equal 1.8×10^{-5} M at 25°C, indicating this to be a weak acid (i.e. it does not dissociate appreciably in solution).

Therefore, the larger the value of K_a the stronger the acid. K_a values, like $[H^+]$ values, are often very small, and so the value is sometimes expressed as the logarithm of its reciprocal, or the negative logarithm, of the value. This is called the **pK_a**. Therefore,

$$pK_a = -\log K_a$$

The smaller the value of pK_a the stronger the acid. For ethanoic acid, pK_a = $-\log (1.8 \times 10^{-5}\ M) = -(-4.74) = 4.74$.

Similarly the dissociation of a weak base, which results in the production of hydroxide ions, is represented by the base dissociation constant, K_b. Consider the reaction of ammonia with water:

$$\underset{\text{(weak base)}}{NH_3(aq)} + H_2O(l) \rightleftharpoons \underset{\text{(conjugate acid)}}{NH_4^+(aq)} + OH^-(aq)$$

The dissociation constant is:

$$K_b = \frac{[NH_4^+{}_{(aq)}][OH^-{}_{(aq)}]}{[NH_{3(aq)}]}$$

The value of K_b for ammonia is 1.75×10^{-5} M.

Again **pK_b = $-\log K_b$.**

So pK_b for ammonia $= -\log(1.75 \times 10^{-5}\ M) = -(-4.76) = 4.76$.

REMINDER

The general case of a weak acid, HA, dissociating in water can be represented by:

$$HA(aq) \rightleftharpoons H^+(aq) + A^-(aq)$$

The acid dissociation constant, K_a, is given as:

$$K_a = \frac{[A^-][H^+]}{[HA]}$$

7.7 Buffers and buffered solutions

The metabolic activity of cells results in the production of acids. For example, lactic acid is produced in muscle tissue. The function of proteins and enzymes requires conditions of constant pH. In particular, the pH of blood must be maintained within narrow limits and even a small variation can be fatal. Fortunately there are mechanisms in the body called **buffer systems** that act to minimise changes in pH.

A buffer solution is a solution which can resist changes in pH after the addition of hydroxonium or hydroxide ions. Buffer solutions consist of mixtures of weak acids or bases and their salts. For example carbonic acid, H_2CO_3, and sodium hydrogen carbonate, $NaHCO_3$; or ammonia, NH_3, and ammonium chloride, NH_4Cl. Such combinations of acids and bases and their salts are called **conjugate acid–base** pairs. Conjugate acid–base pairs are linked by the gain and loss of a proton (or hydoxonium ion).

For example, carbonic acid and the hydrogen carbonate ion, H_2CO_3/HCO_3^-

$$\underset{\text{(conjugate acid)}}{H_2CO_3(aq)} + \underset{\text{(base)}}{H_2O(l)} \rightleftharpoons \underset{\text{(conjugate base)}}{HCO_3^-(aq)} + \underset{\text{(acid)}}{H_3O^+(aq)}$$

ammonia and the ammonium ion, NH_3/NH_4^+

$$\underset{\text{(conjugate base)}}{NH_3(aq)} + \underset{\text{(acid)}}{H_2O(l)} \rightleftharpoons \underset{\text{(conjugate acid)}}{NH_4^+(aq)} + \underset{\text{(base)}}{OH^-(aq)}$$

Note that in each of these dissociations the water molecule and the hydroxonium ion formed are also a conjugate acid–base pair.

A buffer system is one that contains almost equal concentrations of conjugate acid and base.

An acidic buffer is one that will maintain a pH of less than 7. If a small amount of base in the form of the hydroxide ion (OH^-) is added to such a buffer, the conjugate acid will react with the hydroxide ions, thus preventing the rise in pH (to more basic conditions). If a small amount of acid in the form of hydroxonium ion (H_3O^+) is added to the buffer, the conjugate base will react with the H_3O^+ ions and prevent a fall in pH (to more acidic conditions). Such a buffer system can maintain an almost constant pH until either the conjugate acid or base is used up. Thus the higher the concentrations of conjugate acid and base, the greater the buffering capacity of the buffer.

TAKING IT FURTHER: Biological buffers (p. 111)

7.8 Calculating the pH of buffer systems using the Henderson–Hasselbalch equation

Every buffer system that is composed of a specific conjugate acid–base pair has a specific pH which it is able to maintain through its buffering action. Thus, different buffers will be used in different biological and chemical systems. The pH of a particular buffer system can be calculated using the **Henderson–Hasselbalch equation.**

This is derived in the following manner for a general conjugate acid–base pair, HA/A^-.

Consider the dissociation of the weak acid HA in water to its conjugate base, A^-:

$$HA(aq) + H_2O(l) \rightleftharpoons A^-(aq) + H_3O^+(aq)$$

The acid dissociation constant for the weak acid is defined as:

$$K_a = \frac{[A^-][H_3O^+]}{[HA]}$$

This can be written in the following way to show the ratio of conjugate base to acid at equilibrium:

$$K_a = [H_3O^+] \times \frac{[A^-]}{[HA]}$$

If we take logs to the base 10 of all the components of this equation in order to enable us to use the 'pK_a' notation we get:

$$\log K_a = \log [H_3O^+] + \log [A^-] - \log [HA]$$

and multiplying through by minus one (–1):

$$-\log K_a = -\log [H_3O^+] - \log [A^-] + \log [HA]$$

Substituting $pK_a = -\log K_a$ and $pH = -\log [H_3O^+]$ gives:

$$pK_a = pH - \log [A^-] + \log [HA]$$

which can be rearranged to:

$$pH = pK_a + \log \frac{[A^-]}{[HA]}$$

A more convenient way of writing this equation is:

$$\mathbf{pH = pK_a + \log \frac{[base]}{[acid]}} \text{ or } \mathbf{pH = pK_a + \log \frac{[proton\ acceptor]}{[proton\ donor]}}$$

This is known as the **Henderson-Hasselbalch equation.**

By using pK_a values, we are able to express the strength of an acid (i.e. its tendency to dissociate) with reference to the pH scale.

7.9 Life in water

TAKING IT FURTHER: Solubility in water (p. 42)

The solubility of substances in water is determined by their structure. Solids which easily dissolve in water to form ions do so because the charged ions readily interact with the polar water molecules. Similarly, any organic molecule that possesses polar covalent bonds (e.g. the O-H bond in alcohols and carboxylic acids) is likely to be soluble in water.

Common functional groups on biological molecules include the hydroxyl group, the carboxyl group and the amino group (see Section 5.7).

The **hydroxyl group (OH)**, consists simply of a hydrogen atom bonded to an oxygen atom, which in turn is bonded to a carbon atom. The bond between the hydrogen and the oxygen is highly polar and thus this functional group strongly attracts water molecules, forming hydrogen bonds. The high solubility of sugars, for example, is due to the presence of hydroxyl groups. Organic molecules containing hydroxyl groups as the only functional group are called **alcohols**. Under normal biological conditions, the hydroxyl group does not dissociate.

The **carboxyl group (COOH)** consists of a carbon atom which is doubly bonded to an oxygen atom (as in the carbonyl group) and also bound to a hydroxyl group. Molecules containing a carboxyl group as the only functional group are called **carboxylic acids,** and the ethanoic acid found in vinegar is an example. The reason that the carboxyl group has acidic properties (is a proton donor) is that the bond between the oxygen and the hydrogen is so highly polar (more so due to the proximity of the carbon–oxygen double bond) that the hydrogen tends to dissociate from the molecules as a free H^+ ion (combining with water to form the hydroxonium ion) and produces a carboxylate ion (Fig. 68).

$$R{-}C({=}O^{\delta-}){-}O^{\delta-}{-}H^{\delta+} + H_2O \rightleftharpoons R{-}COO^- + H_3O^+$$

carboxylate ion

Figure 68. The dissociation of a carboxyl group

The dissociation constant K_a, for the carboxyl group, can be shown as:

$$K_a = \frac{[RCOO^-][H_3O^+]}{[RCOOH]}$$

The RCOOH acts as a weak acid and water as a base. The carboxylate ion formed can act as a base (accepting H^+) with water acting as an acid.

In forming the **amino group ($-NH_2$)**, nitrogen has atomic number 7 and therefore the electron configuration $1s^2 2s^2 2p^3$. When nitrogen hybridises it has three valence electrons (three hybrid sp^3 orbitals) and one lone pair of electrons.

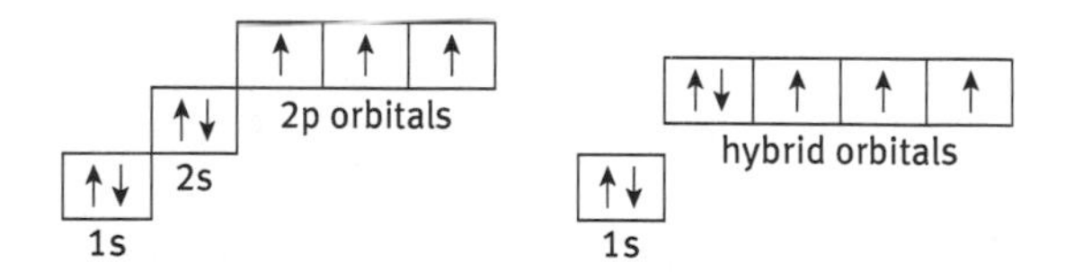

Thus, in the amino group the nitrogen atom possesses a lone pair of electrons. This can readily associate with a proton from water to form $-NH_3^+$, which in turn can also dissociate (Fig. 69). The NH_2 group acts as a base and water as an acid. The $-NH_3^+$ group formed also acts as an acid by dissociating to lose a proton to water, which acts as a base in accepting the proton.

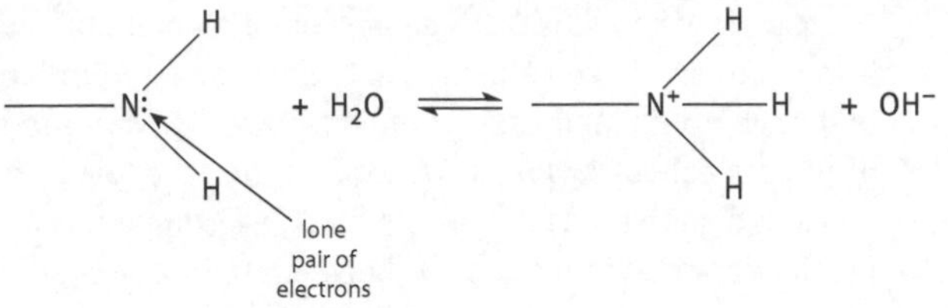

$$-\ddot{N}H_2 + H_2O \rightleftharpoons -N^+H_3 + OH^-$$

lone pair of electrons

Figure 69. Protonation of the amino group

The dissociation of a proton from the protonated form of the amino group of an amino acid can be described in terms of its K_a value.

$$RNH_3^+\ (aq) + H_2O\ (l) \rightleftharpoons RNH_2\ (aq) + H_3O^+\ (aq)$$

$$K_a = \frac{[RNH_2][H_3O^+]}{[RNH_3^+]}$$

7.10 Amino acids

The general structure of an amino acid is shown in Fig. 70. All amino acids possess an amino group and a carboxyl group; these are both charged at physiological pH (pH 7). However, in proteins the amino acids are linked together through the formation of peptide bonds and so these charged groups are effectively 'lost'. However, the 'R' group is specific for each individual amino acid and includes groups which are polar, acidic, basic and apolar (hydrophobic). The R groups of amino acids, and therefore of proteins, are of central importance in determining the nature of intermolecular interactions within protein molecules (which determine their three-dimensional shapes) and between protein molecules and other molecules in the cell (such as between an enzyme and its substrate).

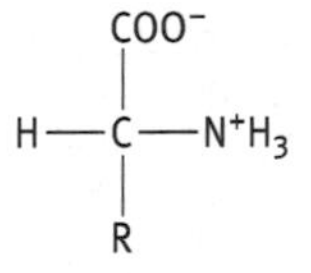

TAKING IT FURTHER:
The peptide bond
(p. 28)

$$H-\underset{R}{\overset{COO^-}{C}}-N^+H_3$$

Figure 70. General structure of an amino acid at physiological pH

Carboxyl and amino groups are commonly found in the side chains of amino acids. The pK_a values of these groups are such that at physiological pH 7, proteins will contain carboxylate groups (COO^-) and protonated amino groups (NH_3^+).

For example, glutamic acid possesses a carboxyl side group (R group). This has a pK_a of 4.07 and so at physiological pH will be dissociated and carry a negative charge.

glutamic acid

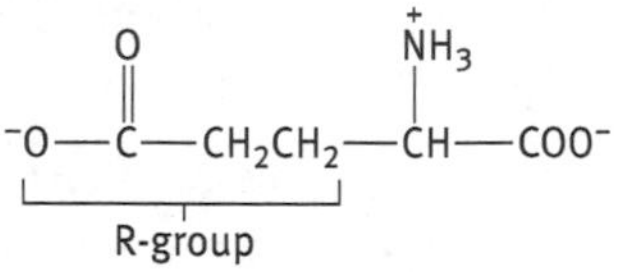

$$^{-}O—\overset{\overset{\displaystyle O}{\|}}{C}—CH_2CH_2—\overset{\overset{\displaystyle \overset{+}{N}H_3}{|}}{CH}—COO^{-}$$

R-group

Lysine possesses an amino side group (R group) with a pK_a of 10.53; this will be protonated at physiological pH and carry a positive charge.

lysine

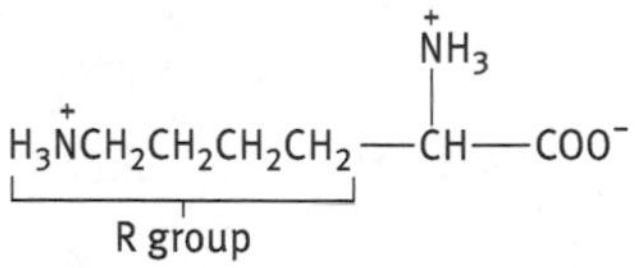

$$H_3\overset{+}{N}CH_2CH_2CH_2CH_2—\overset{\overset{\displaystyle \overset{+}{N}H_3}{|}}{CH}—COO^{-}$$

R group

REMINDER

How to 'read' pK_a. At a pH below the pK_a of a dissociable group, that group will be predominantly protonated. Thus, at physiological pH 7, the carboxyl group will be dissociated (COO^-), and the amino group will be protonated (NH_3^+)

7.11 Controlling cellular pH

TAKING IT FURTHER: Biological buffers (p. 111)

The physiological pH is around neutrality, about pH 7.4. Changes in pH can have radical effects on biological molecules, affecting their structures and activities. Clearly it is important that the organism can control its internal pH; it needs to buffer against pH changes.

A NOTE ON LOGARITHMS

The pH scale is a logarithmic scale representing the concentration of H^+ ions in a solution. Remember, from algebra, that we can write a fraction as a negative exponent, thus 1/10 becomes 10^{-1}. Likewise 1/100 becomes 10^{-2}, 1/1000 becomes 10^{-3}, etc. Logarithms are exponents to which a number (usually 10) has been raised. For example log 10 (pronounced 'the log of 10') = 1 (since 10 may be written as 10^1). The log 1/10 (or 10^{-1}) = −1. pH, a measure of the concentration of H^+ ions, is the negative log of the H^+ ion concentration. If the pH of water is 7, then the concentration of H^+ ions is 10^{-7} M, or 1/10 000 000 M. In the case of strong acids, such as hydrochloric acid (HCl), an acid secreted by the lining of the stomach, $[H^+]$ is 10^{-1} M; therefore the pH is 1.

7.12 Summing up

1. Water is a dipolar molecule which readily undergoes hydrogen bonding with other water molecules.

2. The dissociation, or auto-ionisation, of water forms the hydroxonium ion (H_3O^+) and hydroxide ion (OH^-).

3. The degree of dissociation of water is measured by the equilibrium constant which is the ratio of the product of dissociated ions to the undissociated molecule.

   $$K_w = [H^+][OH^-]$$

4. An acid is defined as a substance which produces hydrogen ions (H^+) by dissociation; bases are defined as substances which can extract a proton, H^+, from the solvent water to leave a hydroxide ion, OH^-.

5. The pH scale is a logarithmic scale in which the pH of a solution is defined by:

   $$pH = -\log[H^+]$$

6. Weak acids are substances which react only partially with water to give hydroxonium ions in solution. A weak base reacts only partially with H_2O to give OH^- ions in solution.

7. The value of the dissociation constant, K_a, for a weak acid is a small number, and therefore is frequently expressed as the pK_a (–log of the dissociation constant).

   $$pK_a = -\log K_a$$

8. $$pH = pK_a + \log\frac{[base]}{[acid]}$$

   This expression is referred to as the Henderson–Hasselbalch equation. It allows us to work out the pH of solutions of buffers, or the pK_a values of conjugate acid and base combinations at a given pH.

9. Carboxyl and amino groups are common functional groups on biological molecules which are dissociated and protonated respectively at physiological pH.

7.13 Test yourself

The answers are given on p. 181.

Question 7.1
Why is it unlikely that two neighbouring water molecules would be arranged like this?

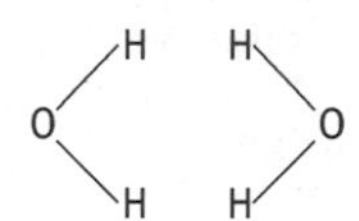

Question 7.2
(a) Give a simple definition of an acid.
(b) What is the difference between a strong acid and a weak acid?
(c) What type of scale is the pH scale?
(d) Compared to a basic solution at pH 9, the same volume of an acidic solution at pH 4 would have how many more hydrogen ions?

Question 7.3
(a) What is the pH of a 0.05 M solution of the strong acid HCl?
(b) What is the $[H^+]$ of a solution of pH 6.2?

Question 7.4
The equilibrium constant (K_a) for the dissociation of ethanoic acid was found to equal 1.8×10^{-5} M. What is the pK_a for ethanoic acid?

Question 7.5
An ethanoic acid and sodium ethanoate buffer system contained 0.10 M of the acid, and 0.05 M of the base. The pK_a of ethanoic acid is 4.75. What is the pH of this system?

Question 7.6
Tris is a weak base that is frequently used to prepare biological buffers. It has a pK_a of 8.08. The pH of the tris base solution is adjusted by addition of HCl. What is the pH of a tris buffer which contains 0.186 M of tris base, and 0.14 M of HCl?

Question 7.7
Molecules X, Y and Z have pK_a values of 4.2, 6.8 and 8.2 respectively. Which is the strongest acid, and what would be the $[H^+]$ in each case? Assume molecules X, Y and Z are present in solution at 1 M concentration.

Taking it further

Biological buffers

Many chemical reactions are affected by the acidity of the solution in which they occur. In order for a particular reaction to occur, or to occur at an appropriate rate, the pH of the reaction medium must be controlled. Biochemical reactions are especially sensitive to pH. Most biological molecules contain groups of atoms which may be charged or neutral, depending on the pH, and whether these groups are charged or neutral, has a significant effect on the biological activity of the molecule.

In all multicellular organisms, the fluid within the cell and the fluids surrounding the cells have a characteristic and nearly constant pH. For example, the pH of the blood in a healthy individual remains remarkably constant at 7.35 to 7.45. This is because the blood contains a number of buffers which protect against pH change caused by acidic or basic metabolites. From a physiological viewpoint, a change of +0.3 or –0.3 pH units is extreme.

This pH is maintained in a number of ways, and one of the most important is through buffer systems. In the laboratory, buffers are typically prepared by combining a solution of a weak acid and a solution of its salt. Take, for example, the ethanoic acid and sodium ethanoate buffer system. In solution, sodium ethanoate ($CH_3COO^-Na^+$) ionises to produce the conjugate base of ethanoic acid, CH_3COO^-.

The equation for the equilibrium that exists in the solution is:

$$CH_3COO^- + H_3O^+ \rightleftharpoons CH_3COOH + H_2O$$

Buffers work by removing H_3O^+ (H^+) or OH^- ions from solution as they are added.

A buffer system obeys Le Chatelier's Principle, which states that:

'When a stress is applied to a system at equilibrium, the system will adjust to relieve the stress'.

Therefore, if hydrogen ions are added they combine with the base to form the conjugate weak acid:

$$\mathbf{H_3O^+} + CH_3COO^- \rightleftharpoons CH_3COO\mathbf{H} + H_2O$$

(the equilibrium above is pushed to the right)

If hydroxide ions are added, the weak acid dissociates to provide H^+ ions, which combine with the OH^- to form H_2O:

$$CH_3COO\mathbf{H} + \mathbf{OH^-} \rightleftharpoons CH_3COO^- + \mathbf{H_2O}$$

(the equilibrium above is pushed to the left)

In either case, the pH does not change dramatically since neither the concentration of H^+ ions nor OH^- ions changes appreciably. Thus, a buffer is able to resist substantial changes of pH.

There are two important biological buffer systems, namely the dihydrogen phosphate system and the carbonic acid system.

1. **The dihydrogen phosphate/hydrogen phosphate system** operates in the internal fluid of all cells. This buffer system consists of dihydrogen phosphate ions ($H_2PO_4^-$) as a hydrogen ion donor (acid), and hydrogen phosphate ions (HPO_4^{2-}) as a hydrogen ion acceptor (base). These two ions are in equilibrium with each other as indicated by the chemical equation below.

   $$H_2PO_4^- + H_2O \rightleftharpoons H_3O^+ + HPO_4^{2-}$$

   If additional hydrogen ions enter the cellular fluid, they are consumed in the reaction with HPO_4^{2-} and the equilibrium shifts to the left. If additional hydroxide ions enter the cellular fluid, they react with $H_2PO_4^-$, producing HPO_4^{2-} and shifting the equilibrium to the right.

   We can see how this system acts as a buffer against pH change, but why does it work so well at neutral pH?

The ability of a compound to act as a buffer at a given pH is determined by how readily it will accept and donate protons (H^+) at that pH. For this reason, any compound which will both accept and donate a lot of protons at a given pH will be an excellent buffer at that pH.

The equilibrium constant expression for this equilibrium is given by:

$$K_a = \frac{[H_3O^+][HPO_4^{2-}]}{[H_2PO_4^-]}$$

and the pK_a (for the dissociation of $H_2PO_4^-$) is found to equal 7.21. In other words, this system is most effective at accepting and/or donating protons at pH 7.21. Buffer solutions are most effective at maintaining a pH near the value of their pK_a. In mammals, cellular fluid has a pH in the range 6.9 to 7.4, and so the dihydrogen phosphate buffer is effective in maintaining this pH range.

2. **The carbonic acid/hydrogen carbonate system:** Another biological fluid in which a buffer plays an important role in maintaining pH is blood plasma. In blood plasma, the carbonic acid/hydrogen carbonate ion equilibrium buffers the pH. In this buffer, carbonic acid (H_2CO_3) is the hydrogen ion donor (acid) and hydrogen carbonate ion (HCO_3^-) is the hydrogen ion acceptor (base).

$$H_2CO_3 + H_2O \rightleftharpoons H_3O^+ + HCO_3^-$$

This buffer functions in the same way as the phosphate buffer. Additional H^+ is consumed by HCO_3^- and additional OH^- is consumed by H_2CO_3. The value of K_a for this equilibrium is 7.9×10^{-7} mol l^{-1}, and the pK_a is 6.1 at body temperature. In blood plasma, the concentration of the hydrogen carbonate ion is about twenty times the concentration of carbonic acid, and therefore provides sufficient buffering capacity to resist excessive changes in the plasma pH [the pH of blood plasma ranges between 7.32 and 7.45].

As we saw earlier, biological molecules contain a number of functional groups that have the capacity to dissociate, notably carboxyl groups and amino groups. So, we might expect proteins (which are polymers of amino acids) to act as buffers, and indeed they do. In fact, albumin (in the plasma) and haemoglobin (in the red blood cell), constitute the largest 'pools' of buffers in the body. As we have stated, compounds act as good buffers at a pH close to their pK_a values. Proteins would be expected to act as buffers at pH values close to the pK_a values of their dissociable amino acid side chains, which for the most part are carboxyl groups and amino groups. Under normal physiological conditions and a pH close to 7, the buffering capacity of such groups is, however, negligible. A look at the table below shows the pK_a of side-chain carboxyl groups to be around 4.0 (for aspartic and glutamic acid), and for amino groups to be around 10 to 13 (for lysine, arginine and asparagine).

However, there is one amino acid whose side chain does have a pK_a value near neutrality, namely histidine.

Amino acid	pK_a of R group
Arginine	Amino – 13.2
Asparagine	Amino – 13.2
Aspartic acid	Carboxyl – 3.65
Glutamic acid	Carboxyl – 4.25
Histidine	5.97
Lysine	Amino – 10.28

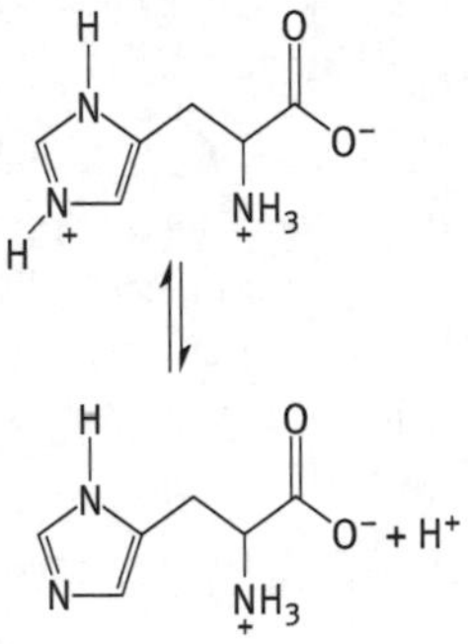

Figure 71. Dissociation of the histidine pyrrole ring nitrogen

At a neutral pH, it is the pyrrole ring nitrogen in the histidine side chain which is in equilibrium,

$$-N^{+}H \rightleftharpoons -N + H^{+}$$

In other words, at neutral pH, it is the histidine in proteins such as albumin and haemoglobin which is responsible for their buffering action.

08 Reacting molecules and energy

BASIC CONCEPTS:

Energy is a central concept in biological systems. Cellular processes are geared to obtaining energy from foodstuffs, and using that energy to do work. Here we consider how the cell obtains energy, how it must overcome 'energy barriers' in order for reactions to occur and the concept of 'free energy' and the laws of thermodynamics.

Energy is required by all organisms for a range of biochemical processes, such as the transport of molecules, the biosynthesis of molecules, the maintenance of pH or osmotic pressure, cellular motility and many others. Animals are chemotrophs; they obtain their energy from breaking down molecules (catabolism). They use much of this energy in the synthesis of new molecules (anabolism). The latter depends on the former, and managing these two processes is the key to 'successful metabolism'.

REMINDER

Both the **calorie** and the **joule** are units of energy. A **calorie** is the amount of energy or heat needed to increase the temperature of one gram of water by 1° Celsius. One calorie has the same energy value as 4.186 joules (J). The nutritional calorie, Cal = 1000 cal = 4.186 kJ

8.1 Energy from molecules

When atoms form molecules through the formation of covalent bonds, energy is released. The energy of the product molecules is therefore less than that of the reactant molecules. It is for this reason that atoms form covalent bonds; to reach a lower energy and therefore a more stable state. For any particular chemical bond, say the covalent bond between hydrogen and oxygen, the amount of energy it takes to break that bond is exactly the same as the amount of energy released when the bond is formed. This value is called the **bond energy** (or bond dissociation energy). Bond energy is the energy required to break a covalent bond homolytically (into neutral fragments). Bond energies are commonly given in units of kcal mol^{-1} or kJ mol^{-1}, and are generally called bond dissociation energies when given for specific bonds, or mean bond energies when summarised for a given type of bond over many kinds of compounds.

REMINDER

The mean **bond energy**, ΔH_B is the average energy change required to break one mole of a given type of chemical bond. e.g. ΔH_B (O-H) = 463 kJ mol^{-1}

Consider the following reaction, the oxidation of methane:

$$CH_4 + 2O_2 \longrightarrow CO_2 + 2H_2O + \textbf{HEAT}$$

Overall this reaction generates heat energy, it is an **exothermic** reaction. Looking at the two sides of the reaction, on the left-hand side of this equation covalent bonds are being broken and energy must necessarily be supplied for this to happen (initially in the form of a spark or flame), therefore this part of the reaction is **endothermic** (energy is absorbed). On the right-hand side of the equation covalent bonds are being formed and therefore energy is released.

REMINDER

Covalent bonds are formed between atoms by sharing electrons; by sharing electrons and filling their outer energy levels, atoms reach a more stable state, i.e. one with a lower energy. It must follow therefore that energy is released when covalent bonds are formed, and absorbed when bonds are broken

Using published tables of bond energies, we can construct an energy balance for this reaction (ΔH_B = bond energy).

ENERGY IN			**ENERGY OUT**		
Energy to break four C-H bonds	$4 \times \Delta H_B$ (C-H)	+1648 kJ	Energy to form two C=O bonds	$2 \times \Delta H_B$ (C=O)	–1486 kJ
Energy to break two O=O bonds	$2 \times \Delta H_B$ (O=O)	+992 kJ	Energy to form four O–H bonds	$4 \times \Delta H_B$ (O–H)	–1852 kJ
		= +2640 kJ			= –3338 kJ
Energy balance =					-698 kJ

Thus, overall the reaction is exothermic and releases 698 kJ of energy per mole of CH_4. Note that a plus sign denotes energy that is put in to a reaction, a minus sign energy that is released from a reaction.

8.2 Getting molecules to react

Getting one molecule to react with another first requires the molecules to collide. The reacting molecules must have enough energy and collide in the correct direction such that one or more covalent bonds are broken. The energy needed to initiate this reaction (i.e. to break one or more covalent bonds) is referred to as the **activation energy, E_a**; it is the minimum amount of energy required for reaction to occur.

REMINDER

The **activation energy**, E_a, of a reaction is the minimum amount of energy required for reaction to occur. It is specific for a specific reaction

The reacting molecules must be raised in energy to a **'transition state'**. In chemistry providing heat to the reaction is one way to raise molecules to the transition state; this increases the number of successful collisions and therefore increases the rate of the reaction. Another way to achieve this would be to provide an alternative way for the reaction to occur, one which has a lower activation energy. The alternative route is provided by a **catalyst**.

REMINDER

A **catalyst** changes the rate of a reaction without being consumed in the reaction. It achieves this by providing an alternative mechanism for the reaction with a different value of E_a

We can summarise this in an 'energy profile' diagram (Fig. 72).

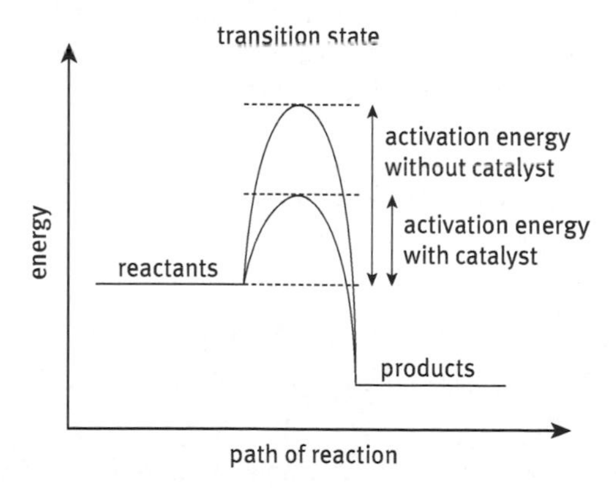

Figure 72. Energy profile diagram

The reactants must be raised by a minimum amount of energy, the activation energy, to reach the transition state in order to initiate the reaction. In the presence of a catalyst the activation energy is smaller and so less energy needs to be supplied to initiate the reaction.

All reactions require an activation energy, whether the overall reaction is exothermic or endothermic. The two energy profiles in Fig. 73 show an exothermic reaction (A) and an endothermic reaction (B).

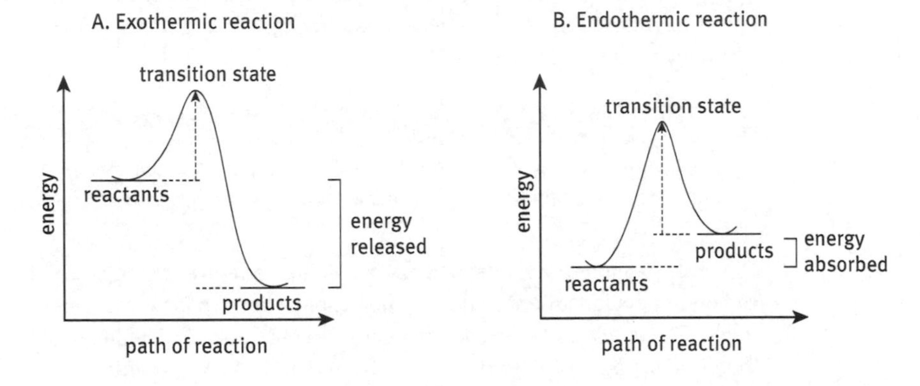

Figure 73. Energy profiles of an exothermic and an endothermic reaction

In reaction A, the energy content of the products is less than that of the reactants; energy must have been released in the reaction, so the reaction is exothermic. In reaction B, the energy content of the products is higher than the reactants so energy must have been absorbed in the reaction; the reaction is endothermic.

> **REMINDER**
>
> An **exothermic** reaction is one in which heat energy is lost to the surroundings. The reaction mixture usually gets warm. An **endothermic** reaction is one in which heat energy is taken in from the surroundings. The reaction mixture usually gets cool

Biological enzymes, like catalysts, lower the activation energy of a reaction by providing an alternative reaction route.

8.3 Energy, heat and work: some basic terms of thermodynamics

All living things require a continuous throughput of energy. Their metabolism ultimately transforms the energy to heat, which is dissipated to the environment. A large portion of the cell's biochemical apparatus is therefore devoted to the acquisition and utilisation of energy. **Thermodynamics** (Greek: *therme* = heat and *dynamis* = power) is the science which describes the relationships between the different forms of energy.

Although complicated, biological systems obey the fundamental laws of thermodynamics.

In thermodynamics we refer to the 'system' (the object being studied) and its 'surroundings' (everything else in the universe). The system is often a cell, but may also be a biological polymer or molecule (the basic thermodynamic laws apply whether the system is 'living' or not).

The first two laws of thermodynamics may be stated as follows:

- The first law states that the total energy of the universe is always conserved. Energy may be converted into different forms but energy is neither created nor destroyed. Energy lost by the system, must be gained by the surroundings and *vice versa.*
- The second law of thermodynamics expresses the phenomenon that the universe tends towards maximum disorder, or in other words, the direction of all spontaneous processes is such as to increase the **entropy** of a system plus its surroundings. This simply reflects our commonsense understanding that things, if left alone, will not get more orderly!

REMINDER

Entropy, *S*, is a measure of the disorder of a system

Such laws may appear fairly abstract but they are obeyed by all physical, chemical and biological systems without exception. How useful they are in a biological context depends on how we use them! In biology we seek to use thermodynamics to tell us how feasible a particular process is, in which direction it is likely to proceed and what the energy requirements are. We measure such changes under *standard conditions* (see Appendix 4).

8.4 Enthalpy

The heat change in a constant-pressure reaction (as is the case in biological processes) is given by the term delta ***H*** (ΔH), the **enthalpy change**. As we have seen above, reactions during which a system releases heat are known as exothermic processes. They have a negative value of ΔH (a negative value is used to show that the reaction is 'losing' heat). ΔH can help us keep track of energy changes and is useful since it depends only on the initial and final states of the system, but it does not indicate the favoured direction of a reaction. Many spontaneous reactions are exothermic, but some are endothermic.

REMINDER

The **enthalpy change** in a reaction is equivalent to the heat change at constant pressure. It is measured in joules per mol (J mol^{-1})

REMINDER

The Greek letter 'delta' (Δ) is used to denote 'a change in'. We cannot measure absolute values of enthalpy or entropy, but we can measure changes in them

8.5 Entropy

The **entropy, *S***, of a system is a measure of a system's degree of disorder; an increase in entropy is described by a positive ΔS. This is a statement of the second law of thermodynamics. All spontaneous reactions result in an increase in entropy. Entropy is a powerful indicator of the direction of a process, but in complex biological systems is extremely difficult to measure.

Because living systems exchange energy with their surroundings, both energy and entropy changes will take place, and both are important in determining the direction of thermodynamically favourable processes. All living things are open systems and, as such, exchange energy with their surroundings in two ways:

- by heat transfer
- by doing work on the surroundings (or having work done on them).

8.6 Gibbs free energy and work

Work can take many forms; expansion (e.g. lungs expanding), electrical (e.g. ion movement, nerve impulse), movement of a flagellum, contraction of muscle, etc. We need a thermodynamic function which includes both energy and entropy, and which will tell us how much work can be done. There are a number, but of prime importance in biology is the **Gibbs free energy**. In 1878 J.W. Gibbs devised a formula which combines both the first and second laws of thermodynamics. It introduces the term free energy, abbreviated to ***G*** in honour of Gibbs.

REMINDER

The **Gibbs free energy, *G***, is the amount of energy in a system which is free to do work at constant temperature and pressure

It is defined by the relationship:

$$\Delta G = \Delta H - T\Delta S$$ where T is the absolute temperature.

Biologists use the symbol $\Delta G^{o\prime}$ to denote the standard free energy change under biological conditions (see Appendix 4).

REMINDER

Absolute temperature is measured on the Kelvin scale.
273 K = 0° C; −273° C = 0 K

Processes in which there is a negative enthalpy change (ΔH is negative – i.e. an exothermic reaction) and/or an increase in entropy (ΔS is positive) are typical of favourable reactions, producing a negative ΔG value.

A process with a negative ΔG is an **exergonic** process; free energy is released and may be available to do work.

A process with a positive ΔG is an **endergonic** process; free energy must be provided for the reaction to proceed.

ΔG is of fundamental importance in biology:

- it indicates whether a process will or will not occur
- it indicates in which direction a process will proceed
- it indicates how far from equilibrium a process is
- it indicates how much useful work may be available from a process.

The following three examples show how enthalpy and/or entropy contribute to the direction of chemical reactions. In each case the calculated value for ΔG is negative (the reaction is exergonic). We can conclude therefore that in each case the reaction is thermodynamically feasible, it will occur in the direction indicated and that the free energy released might be available to do work.

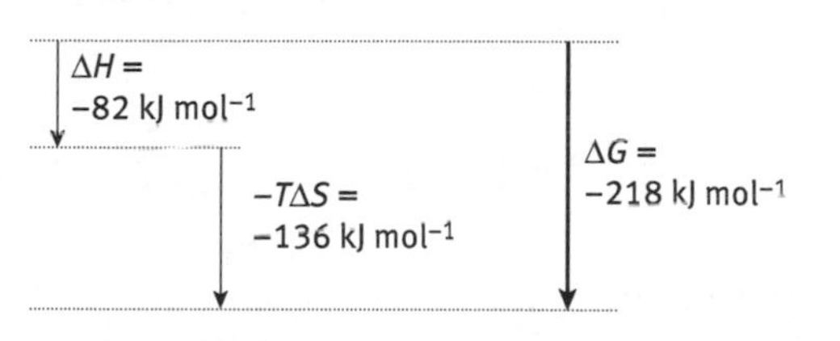

$C_6H_{12}O_6$ (s) ⟶
glucose

$2C_2H_5OH$ (l) + $2CO_2$ (g) + $3H_2O$ (l)
ethanol

Enthalpy (ΔH) and entropy ($T\Delta S$) changes both favour this reaction

Fermentation of glucose to ethanol

$\Delta G = \Delta H - T\Delta S = (-82 - 136)$ kJ mol^{-1} = -218 kJ mol^{-1}

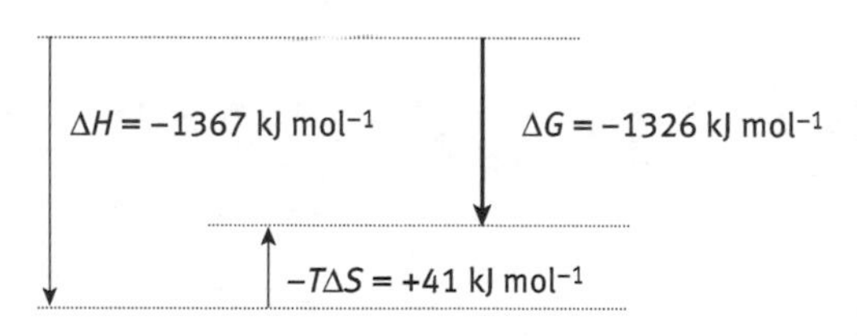

C_2H_5OH (l) + $3O_2$ (g) ⟶
$2CO_2$ (g) + $3H_2O$ (l)

The enthalpy change (ΔH) favours this reaction (it is 'enthalpy-driven'). The entropy change is slightly positive. If H_2O (g), i.e. water vapour, were the product, then the reaction would also be entropy-driven.

Oxidation of ethanol

$\Delta G = \Delta H - T\Delta S = (-1367 - (-41))$ kJ mol^{-1} = -1326 kJ mol^{-1}

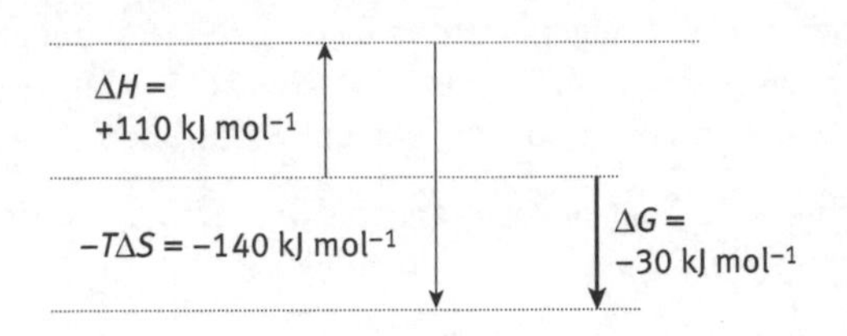

$$N_2O_5\,(s) \longrightarrow 2NO_2\,(g) + {}^1\!/_2O_2\,(g)$$

The reaction actually absorbs heat (ΔH is positive), but there is a large entropy increase (it is 'entropy-driven') due to the formation of gaseous products (more disorder).

Decomposition of dinitrogen pentoxide

$$\Delta G = \Delta H - T\Delta S = (110 - 140)\text{ kJ mol}^{-1} = -30\text{ kJ mol}^{-1}$$

Biological processes for which ΔG is positive are 'energy-consuming' or endergonic. They take place only when **coupled** to a strongly exergonic process. The cell has to do work to render the process exergonic overall. Such coupling mechanisms are the key to understanding the many processes of metabolism. Energy released from the breakdown of biomolecules is used to provide the energy for the biosynthesis of others. As we stated at the beginning of this chapter, anabolism depends on catabolism, and the management of metabolism depends on the successful integration of the two.

8.7 Energy changes in biological reactions

Consider one further exothermic reaction, the oxidation of hydrogen. If you ignite a mixture of hydrogen and oxygen (the spark provides the necessary activation energy) the result is a dramatic explosion. The equation for this chemical reaction is:

$$2H_2\,(g) + O_2(g) \rightarrow 2H_2O(l)$$

And, as the explosion suggests, a release of energy occurs. In fact, the energy change is -447 kJ mol^{-1}.

REMINDER

The cell has to do work, which it achieves by coupling an exergonic process with an endergonic process

Where does this energy go? In this case it is lost as heat and sound. But biological systems have learnt to control such processes such that the energy released can be usefully used. This chemical reaction may not seem very 'biological', but in fact it is a good model for the reaction at the very heart of life itself.

The subcellular organelles, mitochondria, exploit a similar reaction to the above to secure ‘free energy’. Mitochondria synthesise water using the hydrogen atoms removed from organic molecules like glucose, and the oxygen atoms they take in as they respire. The process is called **cellular respiration**. Cellular respiration results in a large negative ΔG.

The overall equation is:

$$C_6H_{12}O_6 + 6O_2 \rightarrow 6CO_2 + 6H_2O \qquad \text{with the release of 2875 kJ mol}^{-1}$$

This is energy which can be harnessed to do work. Mitochondria release this ΔG in small steps which they use to ‘drive’ an endergonic reaction, namely the synthesis of ATP (adenosine triphosphate). ATP is the ‘energy currency’ of the cell; its subsequent breakdown releases free energy which can in turn be used to drive other endergonic reactions. This is a major strategy by which the cell uses free energy derived from catabolism, to drive often unfavourable anabolic reactions. The success of the mitochondrion resides in the fact that released free energy is used to do work, rather than simply being lost as heat, as is the case in the equivalent chemical reaction. This strategy is further developed in ‘Taking it Further: Free energy and metabolic pathways’ and in Chapter 10.

We have seen that energy is required (bond energy) to break covalent bonds, that some reactions can result in a net energy release (exothermic and exergonic reactions), and that a portion of this energy released (free energy) might be used (coupled) to do work and to drive endergonic reactions. The energy changes which occur during the various cellular processes indicate how likely (how thermodynamically feasible) such processes are, in which direction they are likely to proceed and whether they are likely to provide or require energy. This understanding does not allow us, however, to quantify how fast a reaction will proceed, nor to know whether it will go to completion. Indeed it may tell us very little about the molecular mechanism of the reaction. Such information may be gained through the study of **kinetics**, which we consider in Chapter 9.

8.8 Summing up

1. Bond dissociation energy is the amount of energy required when a covalent bond is broken. This amount of energy must be supplied in order to subsequently break that bond. When the same covalent bond is formed, the same amount of energy is released.

2. Chemical reactions may be exothermic (energy is released, ΔH is negative) or endothermic (energy is absorbed, ΔH is positive).

3. Energy must always be supplied to reacting molecules in order to ‘activate’ the reactants; this is referred to as the activation energy, E_a.

4. The activation energy raises the reacting molecules to one or more transition states.
5. A catalyst provides an alternative reaction route with a different, usually lower, activation energy.
6. All biological systems obey the fundamental laws of thermodynamics; 'energy may be converted but is neither created nor destroyed', and 'spontaneous processes result in an increased disorder'.
7. Enthalpy change (ΔH), a measure of heat change, and entropy change (ΔS), a measure of disorder, are related to the Gibbs free energy (ΔG) by the expression $\Delta G = \Delta H - T\Delta S$
8. Reactions may be enthalpy- and/or entropy-driven.
9. ΔG is of fundamental importance in biology; it will indicate whether a process is feasible, its directionality, the extent of the process and whether that process can be used to do work.
10. A negative ΔG is indicative of an exergonic reaction and one which might provide free energy to do work. A positive ΔG is indicative of an endergonic reaction in which energy must be provided in order for the reaction to proceed.
11. Biological processes couple endergonic processes with exergonic processes, free energy from the latter being used to drive the former.

8.9 Test yourself

The answers are given on p. 181.

Question 8.1
Which of the following responses is correct?
Cells cannot use heat to perform work because:
(a) heat is not a form of energy
(b) cells do not have much heat
(c) temperature is usually uniform throughout a cell
(d) heat cannot be used to do work
(e) heat denatures enzymes

Question 8.2
Which of the following processes can occur without a net influx of energy from some other process?
(a) $ADP + P_i \rightarrow ATP + H_2O$
(b) $C_6H_{12}O_6 + 6O_2 \rightarrow 6CO_2 + 6H_2O$
(c) $6CO_2 + 6H_2O \rightarrow C_6H_{12}O_6 + 6O_2$
(d) amino acids $\rightarrow$ proteins
(e) glucose + fructose $\rightarrow$ sucrose

Question 8.3
Which of the following responses is correct?
An enzyme accelerates a metabolic reaction by:
(a) altering the overall free energy change for the reaction
(b) making an endergonic reaction occur spontaneously
(c) lowering the activation energy
(d) pushing the reaction away from equilibrium
(e) making the substrate molecule less stable

Question 8.4
Explain what is meant by exergonic and endergonic reactions.

Question 8.5
In the oxidation of glucose to carbon dioxide and water (glucose + oxygen $\rightarrow$ carbon dioxide + water) the enthalpy change ΔH of the reaction was found to be -2807.8 kJ mol^{-1}, and the free energy change ΔG was equal to -3089.0 kJ mol^{-1}.
(a) Calculate the entropy change ΔS per mole of glucose at 37°C, and
(b) Comment on the thermodynamic feasibility of this reaction.

Taking it further

Free energy and metabolic pathways

An organism's metabolism is geared to extracting energy from foodstuffs (catabolism) and using that energy to do work and drive otherwise unfavourable reactions. Central to this strategy is the molecule **ATP, adenosine triphosphate**, the 'energy currency' of the cell. The structure of ATP is given below, along with that of ADP for comparison.

ATP

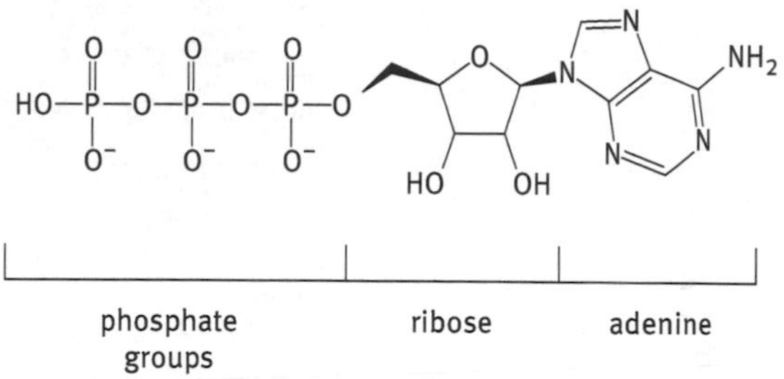

phosphate groups | ribose | adenine

ADP

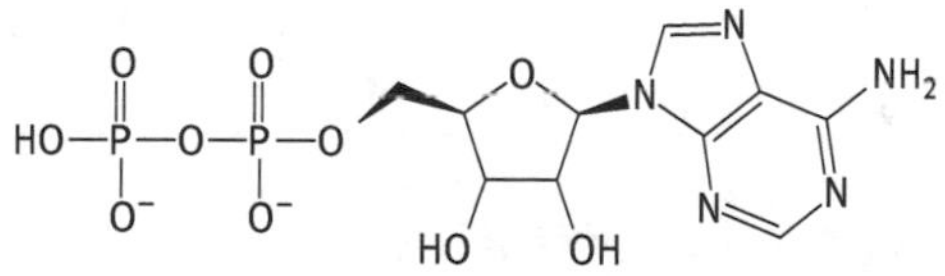

The ATP molecule is constructed from a base, adenine (which we also find in DNA and RNA), linked to a sugar, ribose (which we also find in RNA), which in turn is linked to a series of three phosphate groups. It is the presence of these phosphate groups which is of central importance to the role of ATP. Thermodynamically speaking, ATP is a rather unstable molecule. The three phosphate groups each carry a negative charge at physiological pH, and so there is a lot of repulsion between like charges. Removal of one (or two) of these phosphates alleviates this repulsion and produces a more stable molecule. Enzyme-catalysed hydrolysis of ATP (hydrolysis is the addition of water) yields ADP (adenosine diphosphate) by the removal of one phosphate group.

$$\text{ATP} + \text{H}_2\text{O} \rightarrow \text{ADP} + \text{P}_\text{i} \text{ (inorganic phosphate)}$$

This reaction is spontaneous and highly exergonic; it is associated with a large negative free energy change. The free energy change for this reaction is about 30.7 kJ mol^{-1}. Organisms **couple** the hydrolysis of ATP to drive the energy-consuming activities of the cell. ATP is not the only molecule in the cell which can provide such energy, but it is the main one; evolution has favoured those enzymes which can bind ATP and use its hydrolysis to drive endergonic reactions.

ATP powers most energy-consuming activities of the cell	
Activity	Examples
Anabolic reactions	Synthesis of proteins Synthesis of nucleic acids Synthesis of polysaccharides Synthesis of fats
Active transport of molecules	Transport of molecules and ions across biological membranes
Nerve impulses	
Maintenance of cell volume	Maintenance of osmotic gradients
Addition of phosphate groups to molecules	Phosphorylation of proteins alters their activity
Muscle contraction	
Cell motility	Movement of cilia, flagella and sperm
Bioluminescence	

Metabolic pathways

All metabolic pathways comprise a sequence of serial conversions, from initial reactant(s) through intermediates to final product(s). Intermediates may themselves participate in other metabolic pathways. There are some important concepts which become clear when we consider the many different metabolic pathways in an organism.

- All metabolic pathways are **unidirectional**. There may be individual reactions in the pathway which are clearly reversible, but as a whole the pathway proceeds in one direction only and results in a **net formation** of product.
- The net free energy change of any metabolic pathway is negative. In other words, all metabolic pathways are thermodynamically feasible under physiological conditions and proceed in a particular direction.

This latter point requires some further thought! Catabolic pathways are designed to extract and make available free energy from foodstuffs; they are necessarily associated with large negative free energy changes and by definition are feasible and unidirectional. But what of anabolic pathways? Anabolic processes are unfavourable and energy-consuming, so how can an anabolic pathway be thermodynamically feasible and unidirectional? The answer to this, which we will investigate further, is explained by the coupling of highly exergonic reactions (e.g. the hydrolysis of ATP) to unfavourable endergonic reactions. Of course, not all the reaction sequences of an anabolic pathway are endergonic, and with the input of free energy the overall free energy 'sum' of the whole metabolic pathway is made negative and therefore feasible.

There is one further point to consider.

- Although some metabolic pathways are catalysed by the same enzymes facilitating both the 'forward' (degradation) and 'backward' (synthesis) reactions, organisms always use two separate, non-reversible pathways, one for degradation and one for biosynthesis. This strategy allows the cell to exert metabolic control and balance its energy requirements.

We can consider the above points with a detailed look at two cellular metabolic pathways, namely **glycolysis** and **gluconeogenesis**.

Glycolysis and gluconeogenesis

Glycolysis is that catabolic sequence of reactions which begins the oxidative degradation of glucose. The glycolytic pathway is unidirectional and strongly exergonic, and results in the production of two molecules of ATP for every one molecule of glucose degraded. It is an essential component of the organism's overall energy maintenance; for example, skeletal muscles derive most of their energy (in the form of ATP) from glycolysis. However, organisms also need to synthesise glucose, which can then be stored (as glycogen) in order to provide a constant energy supply for when the body is not ingesting foodstuffs (the brain, for example, needs a constant supply of glucose). The synthesis of glucose is an anabolic process which occurs through the gluconeogenesis pathway ('gluconeogenesis' literally means 'synthesis of new glucose').

Glycolysis

The metabolic pathway for glycolysis is shown in Fig. 74.

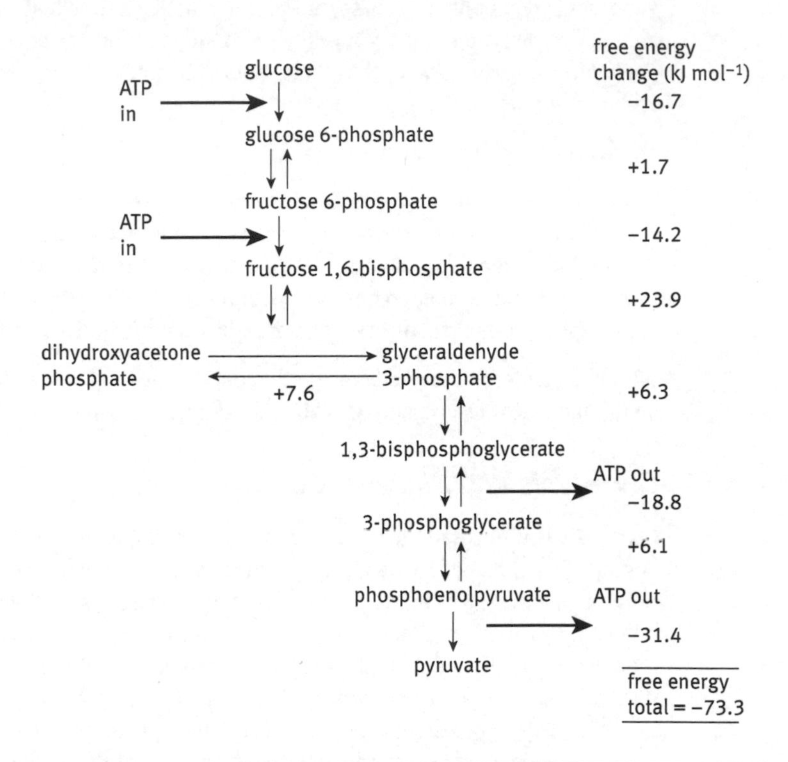

Figure 74. Free energy changes in the glycolytic pathway

Firstly, and in common with most metabolic pathways whether catabolic or anabolic, the pathway has to be 'activated'. In glycolysis this involves the phosphorylation of glucose to glucose 6-phosphate. This reaction requires the input of energy and so is coupled to the hydrolysis of ATP; free energy from ATP hydrolysis drives this reaction forward (ATP_{in}). The reaction becomes unidirectional and is associated with a large negative free energy change. Such irreversible reactions are often referred to as a **'committed step'**. A second input of energy is required to convert fructose 6-phosphate to fructose 1,6-bisphosphate; again the reaction is unidirectional and a large negative free energy change is associated with hydrolysis of ATP. For catabolic pathways such as glycolysis, we can consider an **energy investment phase** and an **energy payoff phase**. In other words energy has to be supplied, to 'kick-start' the pathway, in order that energy can subsequently be derived. In glycolysis, two moles of ATP per mole of glucose are required to move the pathway forward, while four moles of ATP are recovered in two later exergonic reactions (one molecule of glucose provides two molecules of glyceraldehyde 3-phosphate). Thus, the net budget is an energy gain.

Although the more exergonic sequences in this pathway tend to be very unidirectional, there are clearly a number of reactions which are essentially reversible. One might ask the question 'why doesn't the pathway stop or even reverse?' The pathway cannot reverse, at least not completely, because of the exergonic unidirectional component reactions, and the reversible reactions are continuously displaced from equilibrium by the enzymes which rapidly remove the products and so 'pull' those reactions forward.

Gluconeogenesis

The metabolic pathway for gluconeogenesis is shown in Fig. 75. The pathway is shown side by side with glycolysis. Comparing gluconeogenesis with glycolysis, many of the reactions are common to the two pathways (shown by a → symbol) , whereas certain reactions are unique to gluconeogenesis (shown by the •— symbol).

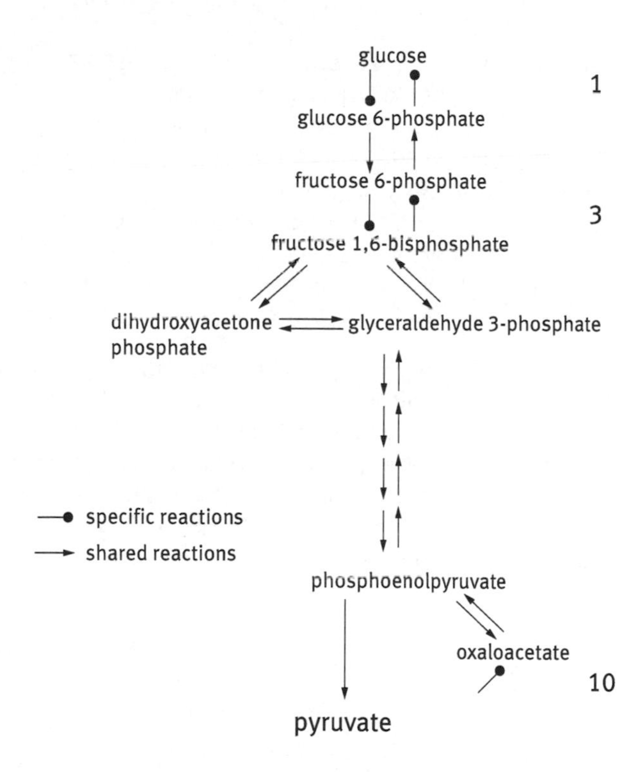

Figure 75. Gluconeogenesis and glycolysis compared

The three unique gluconeogenic reactions, **10**, **3** and **1** in Fig. 75, and emphasised by the 'loops' in Fig. 76, correspond to those highly exergonic steps in glycolysis; these three glycolytic steps cannot simply be reversed and so must be 'by-passed' by other reactions.

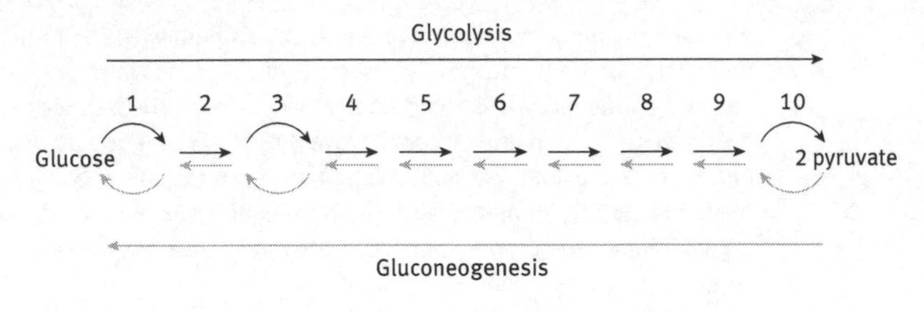

Figure 76. Reaction sequences compared in glycolysis and gluconeogenesis

In gluconeogenesis these ‘by-passes’ are achieved by unique reactions, each requiring an input of energy in order to make them exergonic and unidirectional. The other reversible reactions are shared, although in fact the two metabolic pathways are actually separated within the cell by virtue of occurring in different cellular compartments.

The overall ΔG for gluconeogenesis is negative (–47.6 kJ mol^{-1}) thanks to an input of energy. Glycolysis too has an overall negative ΔG (–73.3 kJ mol^{-1}). The difference is that glycolysis accomplishes a negative ΔG whilst also yielding a net production of ATP, whereas the biosynthetic gluconeogenesis pathway requires an input of energy (partly through ATP hydrolysis) to achieve its overall negative ΔG. One pathway is catabolic and an energy producer (glycolysis) and the other pathway is anabolic and an energy consumer (gluconeogenesis). Energy, provided in the formation of ATP by catabolic pathways, is used to drive anabolic pathways.

The strategies seen in the consideration of glycolysis and gluconeogenesis are applicable to all metabolic pathways within the cell. An input of energy is generally required to ‘commit’ the pathway, usually through the free energy released by the hydrolysis of ATP. This ensures that the overall ΔG for the pathway is negative and that the pathway is unidirectional; reversible reactions are displaced from equilibrium, either by the rapid removal of products, or indeed through high concentrations achieved in the reactants.

09 Reacting molecules and kinetics

BASIC CONCEPTS:
The successful management and integration of the cell's vast array of different metabolic reactions depends, to a large part, on the control and rates of the different reactions. The study of kinetics is a powerful tool for understanding the mechanisms and control of enzyme-catalysed reactions. Here we consider factors which affect the rates of reaction, rate-limiting steps in reaction schemes, and equilibrium states and their relationship to free energy changes.

Chemical kinetics is the study of the rate at which chemical reactions occur and the factors which affect rate. For living organisms the rate of reaction is extremely important; a 'successful' metabolic pathway requires that each individual reaction can occur at an optimal rate. Three factors are important in determining the rate of a reaction.

1. **Temperature**. The higher the temperature, the greater the kinetic motion of the molecules, the more energetic their collisions, and therefore the more likely that the activation energy will be reached or exceeded and a reaction will occur. Higher organisms control their temperature and so the direct effect of temperature on biochemical reactions is of lesser consideration.
2. **Catalysts**. Catalysts increase (or decrease) reaction rates while they themselves remain unchanged. Enzymes are biological catalysts. The physiological temperature of higher organisms is a balance; on the one hand it speeds up the rate and increases the feasibility of the enzyme-catalysed reaction, while on the other it is not so high as to cause undue damage to the delicate protein structure of the enzyme molecule.
3. **Concentration**. The more molecules there are present (the reactants), the more collisions which will occur in a given time, and therefore the greater the rate of the reaction (and the more products which will be formed). In biochemical pathways the concentration of reactants and products is especially important since the catalytic rate of many enzymes can be activated or inhibited by the levels of certain metabolites in those pathways (feedback inhibition).

Rates of reaction are generally expressed in terms of a change in concentration with time, in either the reactants or the products. For example, we might express a rate as moles of product used per second, mol s^{-1} or as the change in concentration of moles of product per second, mol dm^{-3} s^{-1}.

An expression for evaluating the rate of a reaction is

$$\text{rate} = \frac{-\Delta[\text{R}]}{\Delta t}$$

where Δ means a 'change in' and [R] means the molar concentration of the reactants. Of course the concentration of the reactants can only decrease with time and, since the rate can only be positive, a minus sign is included in the equation.

If a product in this reaction were a molecule, P, then we could also write a rate for the reaction in terms of the change in concentration of product per second. In this case the rate of reaction would be expressed as:

$$\text{rate} = \frac{\Delta[\text{P}]}{\Delta t}$$

In this case because the product concentration is increasing with time, the sign is positive.

REMINDER

Rates of reaction are generally expressed as a change in concentration with time

9.1 Rate equations

Consider the reaction A + B → C + D. We can write a rate equation, for the rate of decrease in concentration of the reactants, as:

$$\text{rate} = -k[\text{A}]^x[\text{B}]^y$$

or indeed as the rate of increase in the concentration of the products as:

$$\text{rate} = k[\text{C}]^w[\text{D}]^z$$

k is the **rate constant** for the reaction. The rate constant isn't actually a true constant because it varies if you change the temperature of the reaction or add a catalyst. The rate constant is a constant for a given reaction only if all you are changing is the concentration of the reactants. For these equations, the powers of *x* and *y* are called the **orders of reaction** with respect to reactants A and B respectively; the orders of reaction are usually small whole numbers 0, 1 or 2. The **overall** order of the reaction would be given by x + y.

REMINDER

A rate constant is only a 'true' constant if all that is changing is the concentration of the reactants

In a **zero order reaction** the reaction rate does not depend on the concentration of the reactants and so the order of reaction is zero; any number raised to the power of zero is one, and so the rate of such a reaction would be simply given as:

$$\text{rate} = -k$$

Zero order reactions are actually very common in biology. Where an enzyme is effectively saturated with substrate and is working at its maximal rate, it makes little difference to the rate of the reaction whether the concentration of the reactant (substrate) is slightly raised or lowered; the rate is determined only by the efficiency of the enzyme.

For a **first order reaction** the rate of reaction would depend on the concentration of just one of the reactants. We would write:

Rate $= -k[A]$, if the reaction was first order with respect to A only.

So, if we doubled the concentration of A, then we would double the rate of the reaction.

For a **second order reaction** there are two possibilities to consider. In the first case the rate of reaction may depend only on the concentration of one reactant squared (to the power 2). In this case if the concentration of the reactant is doubled, then the rate will increase 4-fold. The rate equation for a reaction which is second order with respect to A can be written as:

$$\text{rate}' = -k[A]^2 \qquad \text{(a)}$$

Therefore if [A] is doubled to [2A] the new rate, rate″ becomes:

$$\text{rate}'' = -k[2A]^2 = -k4[A]^2$$

If we compare the new rate, rate″, to the old rate, rate′, by dividing one by the other then:

$$\frac{\text{rate}''}{\text{rate}'} = \frac{-4k[A]^2}{-k[A]^2} = 4$$

Thus it can be seen that the new rate is 4 times the old rate if the concentration of reactant A is doubled.

In the second case (b), the rate of reaction depends on the concentrations of two reactants. In the rate equation:

$$\text{rate} = -k[A][B] \qquad \text{(b)}$$

doubling the concentration of either A or B will double the rate, and doubling both concentrations will increase the rate 4-fold.

In these rate equations:

- rate $= k[A]^1$ indicates the reaction is first order with respect to A (although we don't normally show the superscript '1')
- rate $= k[A]^2$ indicates the reaction is second order with respect to A

- rate = k[A][B] indicates the reaction is first order with respect to both A and B, and second order overall.

Orders of reaction can be found only by experiment. The order of a chemical reaction gives you information about which concentrations affect the rate of the reaction. You cannot look at an equation for a reaction and deduce what the order of the reaction is going to be – you have to do some practical work! Having found the order of the reaction experimentally, you *may* be able to make suggestions about the route or mechanism for the reaction, at least in simple cases.

REMINDER

Orders of reaction give information on how the rate is affected by the concentration of the reactants

9.2 Reaction routes or mechanisms

In any chemical change, some bonds are broken and new ones are made. Very often, such changes do not happen in a single step. Instead, the reaction may involve a series of small steps one after the other; this is particularly true of enzyme-catalysed reactions. A reaction mechanism describes the one or more steps involved in the reaction in a way that makes it clear exactly how the various bonds are broken and made.

9.3 The rate-limiting step

Since many reactions occur through several steps, the rate for each step needs to be measured. There will always be one step which is the slowest and that step is called the **rate-limiting step**. The overall rate of a reaction is controlled by the rate of the slowest step. When you measure the rate of an overall reaction, you are actually measuring the rate of the rate-limiting step.

9.4 Considering the activation energy

In Section 8.2 we saw that the reactants must be raised in energy (the activation energy) to a transition state before covalent bonds could be broken and the reaction could proceed. In the diagram for reaction A (Fig. 77), only one transition state is evident, and mechanistically we would assume the reaction to be a relatively simple one.

Reaction A: single transition state

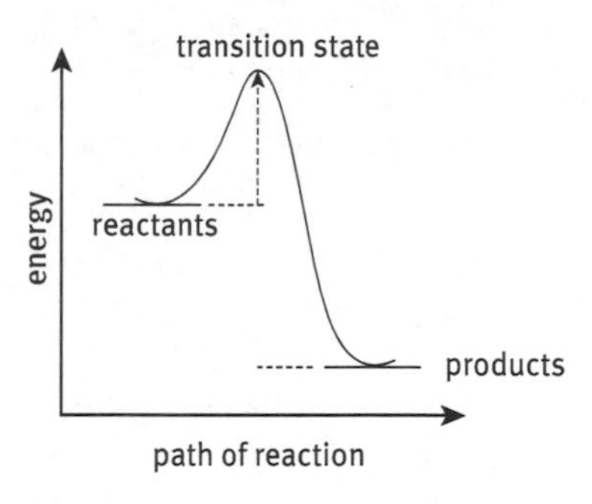

Reaction B: multiple transition states

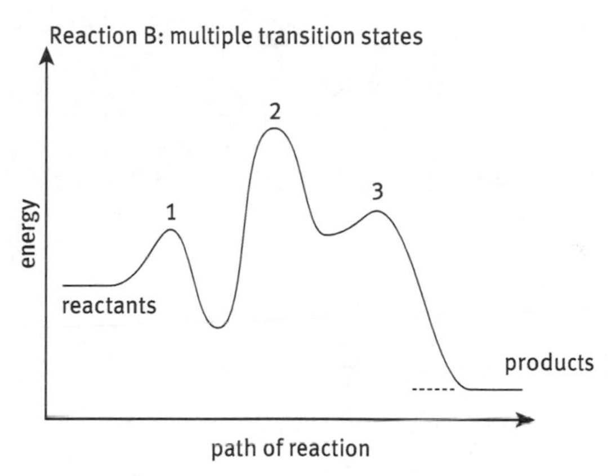

Figure 77. Single and multiple transition states in a reaction

However, generally it is more likely that the reactant molecules would undergo a number of different rearrangements, and involve the breaking and formation of a number of covalent bonds. Each of these rearrangements (transitions) would have a characteristic activation energy. The diagram in reaction B (Fig. 77) shows three transition states. Transition state 2 has the highest activation energy and would therefore represent the rate-limiting step in this reaction.

REMINDER

The rate-limiting step in an enzyme-catalysed reaction will involve the formation of the transition state with the highest activation energy

9.5 Equilibrium

The majority of chemical reactions, and certainly most biochemical reactions, do not proceed to completion. Instead a point is reached at which the products react together to re-form the reactants! The reaction is said to be

reversible. At the point where the rate of the forward reaction (formation of product) equals the rate of the backward reaction (formation of reactant), the reaction is said to have reached **equilibrium**.

In a simple reaction, say the conversion of A to B, we can show the reaction as:

$$A \underset{k^{-1}}{\overset{k^1}{\rightleftharpoons}} B$$

where k^1 is the forward rate (formation of product) and k^{-1} is the backward rate (formation of reactant).

The rate of the forward reaction (formation of products) can be expressed as:

$$\text{rate f} = \frac{\Delta[B]}{\Delta t} = k^1[A]$$

and the rate of the backward reaction (formation of reactants) can be expressed as:

$$\text{rate b} = \frac{\Delta[A]}{\Delta t} = k^{-1}[B]$$

So, because rate f must equal rate b

$$\frac{\Delta[B]}{\Delta t} = \frac{\Delta[A]}{\Delta t}$$

therefore: $k^1[A] = k^{-1}[B]$

so: $\frac{k^1}{k^{-1}} = \frac{[B]}{[A]} = K_c$

where K_c is the equilibrium constant.

REMINDER

At equilibrium, the forward rate (k^1) is equal to the backward rate (k^{-1})

Equilibrium constants, K, may have many different notations depending on the type of equilibrium being investigated (for example, K_w is the equilibrium constant for the ionisation of water). The symbol K_c specifies only that the equilibrium constant shall have some nominal unit. The most common concentration unit is mol per litre or mol l^{-1}. Biologists tend to use K_{eq} for the equilibrium constant of enzyme-catalysed reactions.

For the reduction of pyruvate, which occurs in skeletal muscle, to produce lactate,

$$\text{pyruvate} + \text{NADH} + H^+ \rightleftharpoons \text{lactate} + NAD^+$$

the equilibrium constant K_{eq} can be expressed as:

$$K_{eq} = \frac{[\text{lactate}][\text{NAD}^+]}{[\text{pyruvate}][\text{NADH}][\text{H}^+]}$$

If K_{eq} is very small, then the reaction would lie to the left with little product having been made.

To determine the amount of each compound which will be present at equilibrium you must know the equilibrium constant. To determine the equilibrium constant you first need to 'balance' the equation. Consider the generic equation:

$$\text{aA} + \text{bB} \rightleftharpoons \text{cC} + \text{dD}$$

The upper case letters are the reactants and products. The lower case letters are the coefficients which balance the equation.

The equilibrium constant, K_{eq}, for this general reaction would be expressed as:

$$\frac{[\text{C}]^c\,[\text{D}]^d}{[\text{A}]^a\,[\text{B}]^b}$$

For example,

$$\text{N}_2(g) + \text{H}_2(g) \rightleftharpoons \text{NH}_3(g)$$

This equation as written cannot obey the conservation of mass principle; it does not 'balance'. This is because there are different numbers of each type of atom on each side of the equation (there are 2 nitrogen atoms on the left, but 1 on the right; there are 2 hydrogen atoms on the left, but 3 on the right). The way we balance equations is with 'stoichiometric' coefficients, i.e. numbers in front of the formulae which denote how many 'items' or molecules of that chemical are involved.

Balancing this equation gives:

$$\text{N}_2(g) + 3\text{H}_2(g) \rightleftharpoons 2\text{NH}_3(g)$$

This equation now follows the law of conservation of mass. There are exactly the same numbers of atoms of each element on both sides of the equation. **Stoichiometry** refers to all quantitative aspects of chemical composition and reactions. The equilibrium expression for this reaction would therefore be written as:

$$K_c = \frac{[\text{NH}_3]^2}{[\text{N}_2][\text{H}_2]^3}$$

For example, if one litre of this reaction mixture at equilibrium was found to contain 1.60 moles NH_3, 1.20 moles H_2 and 0.80 moles N_2, then the equilibrium constant would be calculated as:

$$K_c = \frac{(1.60\ \text{M})^2}{(0.8\ \text{M})(1.20\ \text{M})^3} = 1.67\ \text{M}^{-2}$$

The unit for K depends on the units used for concentration. In the above example, concentrations are given as M (mol l^{-1}), and so K has a unit of $M^{2-(4)} = M^{-2}$.

Qualitatively we may look at how far an equilibrium lies towards the right (towards products) or left (towards reactants). The magnitude of the equilibrium constant gives us a general idea of whether the equilibrium favours products or reactants. If the reactants are favoured, then the denominator term of the equilibrium expression will be larger than the numerator term, and the equilibrium constant will be less than unity. If the products are favoured over the reactants, then the numerator term will be larger than the denominator term, and the equilibrium constant will be greater than unity.

9.6 The equilibrium position can change

An equilibrium is able to shift in either direction, either towards reactants or towards products, when a stress is applied. This stress may involve increasing or decreasing the amount of a reactant or product, changing the volume or pressure of a gas phase, or changing the temperature of the equilibrium system. The direction in which the equilibrium shifts may be predicted by using **Le Chatelier's Principle**. This states that if a stress is placed on an equilibrium, then the equilibrium will shift in the direction which relieves the stress. For example, in the following reaction, which represents the muscle conversion of pyruvate to lactate, the equilibrium lies very far to the right favouring the production of lactate.

$$\text{Pyruvate} + \text{NADH} + \text{H}^+ \rightleftharpoons \text{Lactate} + \text{NAD}^+$$

We can 'force' this reaction to the left (pyruvate production) by adding high concentrations of lactate, or indeed NAD^+. In muscle, the enzyme which catalyses this reaction is called lactate dehydrogenase; it is structurally designed to favour a high concentration of lactate (the enzyme has a low affinity for lactate). Lactate, from muscle, is distributed to the liver where it is converted back to pyruvate; here a structurally distinct lactate dehydrogenase favours the production of pyruvate (it has a higher affinity for lactate). This strategy is common in biology; structurally distinct forms of an enzyme (**isoenzymes**) promote different equilibrium positions. It is important to note, however, that whereas there may be an infinite number of equilibrium positions for a reaction, there is only one value of the equilibrium constant (at a defined temperature). The specific equilibrium position adopted by a system depends on the initial concentrations. To underline this fact consider the data below.

The table shows the results of three experiments for the reaction

$$N_2(g) + 3H_2(g) \rightleftharpoons 2NH_3(g)$$

Experiment	Initial concentration	Equilibrium concentration	K_{eq}
1	N_2 = 1.00 M	0.921 M	
	H_2 = 1.00 M	0.763 M	
	NH_3 = 0	0.157 M	**6.02×10^{-2} M^{-2}**
2	N_2 = 0	0.399 M	
	H_2 = 0	1.197 M	
	NH_3 = 1.00 M	0.203 M	**6.02×10^{-2} M^{-2}**
3	N_2 = 2.00 M	2.59 M	
	H_2 = 1.00 M	2.77 M	
	NH_3 = 3.00 M	1.82 M	**6.02×10^{-2} M^{-2}**

Equilibrium concentrations were determined for N_2, H_2 and NH_3, starting with three sets of different initial concentrations. Calculating K_{eq} for each, using the equation

$$k_{eq} = \frac{[NH_3]^2}{[N_2][H_2]^3}$$

gave the same value in each case.

9.7 Free energy and equilibrium

In Chapter 8 we introduced the concept of free energy. Such concepts may appear somewhat abstract. However, by a mathematical combination of the equations for equilibrium constants and those for free energy, we can derive the equation:

$$\Delta G^{o\prime} = -RT \ln K_{eq}$$

(R is a constant called the Universal Gas Constant and T the absolute temperature)

REMINDER

The equilibrium position can change, but the equilibrium constant is fixed at a constant temperature

In other words, the free energy change of a reaction is related to the equilibrium constant for that reaction. If we can measure the equilibrium constant of a biochemical reaction (which is a relatively straightforward procedure) then we can also calculate the free energy change (and *vice versa*). This can provide some very useful information, i.e. whether the process is feasible (likely to occur), in which direction it will proceed, and to what extent it will proceed. These are powerful data in trying to decipher the intricacies and interrelationships of cellular processes.

9.8 Free energy change is zero at equilibrium

Only when a reaction is moving towards equilibrium can there be a net free energy change. For example, for the reaction, A $\rightleftharpoons$ B, there may be a negative ΔG as the reaction moves towards B (the reaction is exergonic), or alternatively ΔG may be positive if the reaction is endergonic (and energy is being supplied to 'drive' the reaction). However, as soon as the reaction reaches equilibrium,

$$\mathrm{A} \underset{k^{-1}}{\overset{k^{1}}{\rightleftharpoons}} \mathrm{B}$$

TAKING IT FURTHER:
Free energy and metabolic pathways
(p. 125)

and the rate of production of B by the forward reaction equals the rate of production of A by the backward reaction, then by definition the net free energy change must be zero. Cellular reactions which are exergonic and generate a negative free energy change generally proceed in one direction only and are displaced from equilibrium by the rapid removal of products.

REMINDER

At equilibrium, rate k^1 = rate k^{-1}, and ΔG is zero

If a cell's metabolic processes reach equilibrium, then that cell is effectively 'dead'! No free energy is produced to do work and there can be no **net** production of products. This is analogous to a battery which has reached equilibrium, i.e. a 'flat' battery; a battery at equilibrium cannot deliver an electromotive force and no current will flow.

9.9 Transport mechanisms

Movement of fluids and solutes within the body depends upon both macroscopic forces (e.g. the pressure within a blood vessel) and microscopic forces that control the movement of individual molecules as they cross cell membranes. Biological cells are characterised by their phospholipid membranes; by their plasma-membrane, that forms a selective barrier with their external environment, and their internal membrane structures that compartmentalise and separate various internal metabolic activities. To varying degrees, biological membranes are semi-permeable; they may be permeable to gases, non-polar or uncharged molecules, or small polar molecules such as water, ethanol and ether, but are impermeable to large polar molecules such as glucose, ions and charged polar molecules such as amino acids and nucleotides. Nevertheless, cells need to import water-soluble nutrients (e.g. sugars, amino acids) and export waste products, as well as control the internal ion concentrations (e.g. H^+, Na^+, K^+, Ca^{2+}). The transport mechanisms employed by biological systems may be simple, varied or highly specific; they may be 'energy-independent' (passive) or energy-dependent (active).

Passive transport systems include:

- diffusion
- osmosis
- facilitated diffusion
- filtration.

Passive transport does not require chemical energy, but is entropy-driven as the process moves towards greater disorder (see Chapter 8). Molecular diffusion, usually referred to simply as **diffusion**, is the net thermal movement of material from an area of high concentration to an area with lower concentration, that may or may not be separated by a semi-permeable membrane. The rate of diffusion is a function of temperature, viscosity of the fluid, size (mass) of the particles, and permeability of the membrane. The difference in concentration between two areas is referred to as the **concentration gradient**. It is important to note that diffusion occurs even in the absence of a concentration gradient; diffusion equilibrium is reached when the concentration of diffusing substances is equal across two compartments.

Osmosis is a special form of diffusion in which the molecules diffusing are water (solvent); water will cross the semi-permeable membrane from a region of low solute concentration (a hypotonic medium) to a region of high solute concentration (a hypertonic medium), in an attempt to equalise the solute concentration on either side. This transport can be countered by increasing the pressure of the hypertonic medium with respect to the hypotonic; thus the **osmotic pressure** is defined as the pressure required to maintain an equilibrium that results in no net movement of water (solvent).

Facilitated diffusion is a passive process like diffusion, but here the movement of molecules across a membrane, down their concentration gradient, is facilitated by a protein carrier, or channel, that is incorporated into the membrane and which allows a molecule to cross an otherwise impermeable membrane.

Filtration refers to the movement of water or solute molecules across a membrane driven by the hydrostatic pressure generated by the cardiovascular system (e.g. the kidney nephron filtration system). Some membranes will only allow very small solutes to pass through, while others allow a wider variety of solutes to pass.

Active transport systems move molecules across biological membranes against their concentration gradients and require the input of chemical energy, usually in the form of adenosine triphosphate (ATP; see Chapter 10). Specific transport proteins in the membrane act as 'pumps', transporting polar and/or charged molecules.

TAKING IT FURTHER: Biological transporters (p. 143)

Both passive and active transport processes require that the overall change in free energy, ΔG, be less than zero, i.e. a negative ΔG. In passive transport this is achieved by the large increase in entropy; in active transport this is achieved by the chemical hydrolysis of (usually) ATP to provide the free energy necessary to 'drive' the system.

9.10 Summing up

1. Rates of chemical reactions are affected by temperature (of lesser consideration for biological reactions), catalysts (enzymes are biological catalysts) and concentration of reactants. In enzyme-catalysed reactions, the concentration of both reactants and products is important because enzymes are often subject to feedback inhibition by their products.
2. Rates of reaction are generally expressed in terms of a change in concentration with time, in either the reactants or the products, and are expressed using rate equations.
3. Orders of reaction indicate the dependency of reaction rate on the concentration of the reactants; a zero order reaction is independent of reactant concentration, a first order reaction is dependent on the concentration of just one of the reactants.
4. The rate of a reaction depends upon the rate-limiting step, which corresponds to the formation of the transition state with the highest activation energy.
5. A reaction may be **reversible.** At the point where the rate of the forward reaction (formation of products) equals the rate of the backward reaction (formation of reactants), the reaction is said to have reached **equilibrium.**
6. The equilibrium constant of a reaction, $\boldsymbol{K_c}$ or $\boldsymbol{K_{eq}}$, is obtained from the product of the product concentrations (raised to the power of their stoichiometric coefficients) divided by the product of the reactant concentrations (also raised to their stoichiometric coefficients). At a stated temperature, the equilibrium constant for a reaction is independent of initial concentrations; there is only one equilibrium constant for a reaction at a specific temperature, but there may be an infinite number of equilibrium positions.
7. A high value for the equilibrium constant indicates that the reaction lies in favour of the formation of products.
8. Le Chatelier's Principle states that if a stress is placed on an equilibrium then the equilibrium will shift in the direction which relieves the stress.
9. Changes in free energy of a reaction are related to the equilibrium constant for that reaction by the expression $\Delta G^{\circ\prime} = -RT \ln K_{eq}$
10. At equilibrium, a reaction has a zero free energy change.

9.11 Test yourself

The answers are given on pp. 181–182.

Question 9.1
Write the equilibrium expression (K_{eq}) for the following reaction:
isocitrate + $NAD^+ \rightleftharpoons$ α-ketoglutarate + CO_2 + NADH

Question 9.2
If a reaction was found to have an equilibrium constant of 3.18×10^{520}, what would this tell you about the reaction?

Question 9.3
In the following reaction:
fumarate + water $\rightleftharpoons$ malate
the ΔG was found to equal −3.7 kJ mol^{-1}. Calculate the K_{eq} at 37°C for this reaction (take R = 8.314 J K^{-1}mol^{-1}), and comment on the value obtained.

Question 9.4
ATP is hydrolysed to produce ADP and inorganic phosphate, P_i; the ΔG for the reaction was found to equal −13 kJ mol^{-1}. Calculate the equilibrium constant for this reaction at 37°C (Take R as 8.314 J K^{-1} mol^{-1}).

Taking it further

Biological transporters

In order to transport molecules, passively or actively, across phospholipid membranes, cells employ a variety of membrane incorporated proteins; these are generally referred to as **carriers** and **channels**.

Carrier proteins cycle between different conformations (like enzymes) in which a solute binding site is accessible on one side of the membrane or the other (Fig. 78).

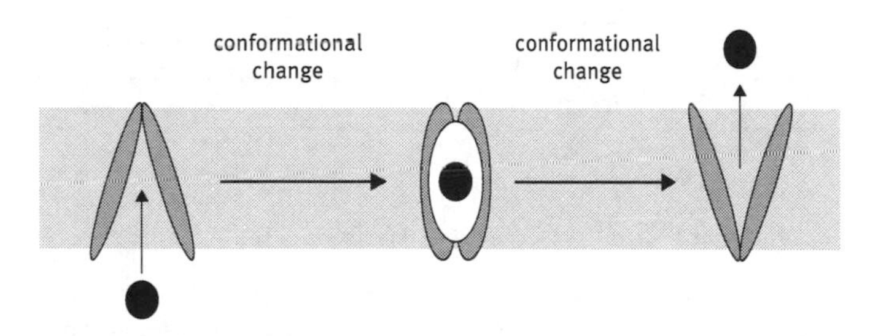

Figure 78. Carrier transporter

Figure 78 shows a solute molecule binding to the protein carrier, which in turn undergoes a number of conformational changes to first internalise then facilitate the release of the solute molecule on the other side of the membrane. Note that at no time is there an ‘open’ channel across the membrane.

Carrier proteins may be referred to as uniport, symport or antiport (Fig. 79).

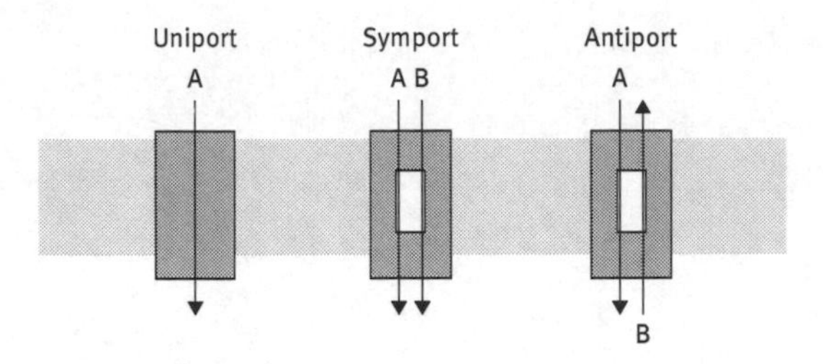

Figure 79. Uniport, symport and antiport carriers

A uniport carrier will transport a single type of solute in one direction across the membrane, a symport carrier will transport two different solutes in the same direction, while an antiport carrier will transport two different solutes in opposite directions across the membrane.

Channel proteins may be 'open' or 'closed'; in their open position they provide a continuous pathway through the membrane through which specific solutes are allowed to pass (Fig. 80).

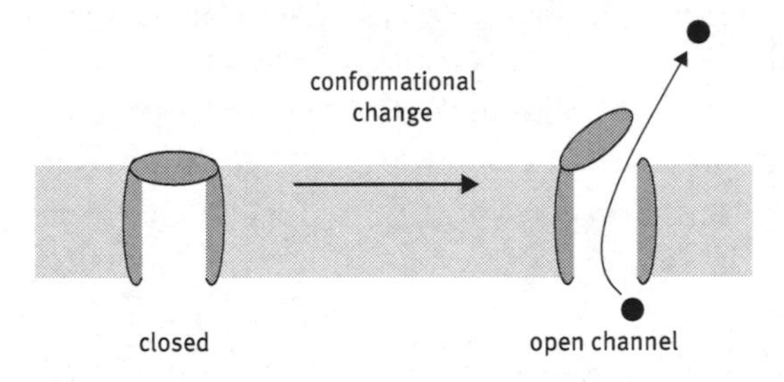

Figure 80. Channel transporter

A conformational change, that may or may not be induced by the solute being transported, is necessary to open the channel 'gate'.

Whether a carrier or channel protein is an active or passive transporter will depend upon the nature of the solute being transported, and whether that solute is being transported down its concentration gradient (passive) or against its concentration gradient (active).

One of the most important active transporters in biological systems is the sodium–potassium pump (Fig. 81). This active transport system, incorporated in the plasma membrane of cells, is responsible for maintaining a relatively low internal sodium concentration and a relatively high internal potassium concentration. It transports both sodium and potassium against their concentration gradients; three sodium ions are 'pumped' out of the cell for every two potassium ions 'pumped' in. An integral part of this protein is an ATP-ase (an enzymic activity responsible for the hydrolysis of ATP, releasing free energy that is used to drive the pump.

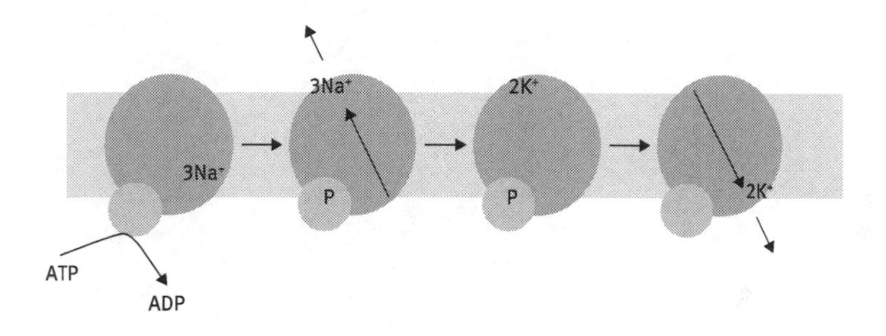

Figure 81. The sodium–potassium pump

ATP binds to the intracellular ATP-ase portion of the protein; in this state three intracellular Na^+ are bound. The ATP is hydrolysed by the ATP-ase; that induces a conformational change in the protein and exposes the Na^+ to the extracellular side where they are released. Two extracellular K^+ are bound and the protein returns to its original conformation, internalising the K^+.

The sodium–potassium pump is particularly important in nerve conduction. The sodium/potassium distribution (the resting potential) across the plasma membrane of neurons can be reversed (causing an action potential); this is the basis of the nerve impulse transmission. In most animal cells, about one-third of the cells' energy expenditure is required for the continuous operation of this pump (up to two-thirds in neurons). In turn, the Na^+ and K^+ gradients set up in the cell can be used to facilitate other transport systems. For example, the Na^+–glucose symporter, which is particularly important in the gut, uses the created Na^+-gradient as a source of energy to import both Na^+ and glucose; this is a much more efficient system than simple diffusion.

10 Energy and life

BASIC CONCEPTS:

The loss or gain of electrons by molecules defines the processes of oxidation and reduction. Life forms use the controlled oxidation of foodstuffs to release free energy, which can be chemically stored or coupled to less favourable reactions to do work. Every oxidation reaction is coupled to a reduction reaction; such redox reactions can be measured by their redox potentials, and in turn be related to the free energy change. Herein lies the very 'driving force' of life and the means by which organisms can build and maintain complex structures in an increasingly disordered environment.

Within the chloroplasts of green plants, light energy is harvested and converted into chemical energy which is stored in ATP and NADPH; these molecules are then used to 'fix' carbon dioxide and produce firstly sugars (carbohydrates), and subsequently amino acids (proteins) and fatty acids (fats). It is estimated that the process of photosynthesis makes some 165 billion tons of carbohydrate a year! Photosynthesis is an endergonic process, an unfavourable process with a positive free energy change and one that is only possible through the supply of energy from the sun (electromagnetic energy). Animals as well as plants then gear their metabolism to the breakdown (catabolism) of organic molecules, releasing free energy (which was originally derived from light) to support the myriad and often unfavourable anabolic processes in the cell. Life would seem to defy the second law of thermodynamics, that the Universe tends towards maximum disorder; our cells are highly ordered and exactly controlled, we are islands of low entropy in an increasingly disordered world! But the laws of thermodynamics are universal. In fact we spend all our time breaking organic molecules down to release energy in order to build others up, but our transfer of energy is far from 100% efficient and we lose much of it as heat, adding to the increasing entropy of the universe. In biological systems:

Energy release = ATP + heat!

Central to life is the ability of an organism to extract and transfer energy from 'food'. A number of strategies have evolved to enable this, the most successful involving the chemical reactions of oxidation and reduction.

10.1 Oxidation and reduction

The original definition of oxidation was 'reaction or combination with oxygen' and reduction was defined as 'reaction or combination with hydrogen'. Our cells oxidise sugars and fats to generate the free energy which we require. The complete oxidation of glucose is shown by the equation:

$$C_6H_{12}O_6 + 6O_2 \rightarrow 6CO_2 + 6H_2O + \text{heat}$$

Both the carbon and hydrogen atoms in glucose have been **oxidised** by bonding to oxygen. In our oxygen-rich atmosphere there is a natural tendency for things to become oxidised (fats go rancid, iron rusts), but a closer look at this equation also shows that the oxygen atoms have been **reduced.** Oxidation of one molecule is always linked to the reduction of another molecule. We speak of oxidation–reduction reactions, which we further abbreviate to **redox** reactions, to emphasise this link.

In biology, redox reactions are defined as those which involve transfer of electrons. All oxidation reactions involve the loss of electrons, and all reduction reactions involve the gain of electrons. The loss or gain of electrons is a better definition of oxidation and reduction since not all redox reactions necessarily involve oxygen or hydrogen. It can, however, be rather more difficult to decide whether electrons are being lost or gained in a reaction. In a simple case, the reaction between sodium and chlorine to form sodium chloride can be written as:

$$\mathbf{Na + \tfrac{1}{2}Cl_2 \rightarrow Na^+\,Cl^-}$$

REMINDER

Oxidation **I**s **L**oss of electrons. **R**eduction **I**s **G**ain of electrons. The mnemonic **'OIL RIG'** can be helpful in remembering this

The strongly ionic compound sodium chloride is formed as a result of sodium losing an electron (being oxidised) and chlorine gaining an electron (being reduced). Sodium, the electron donor, is the **reducing agent** because it reduces the chlorine. Chlorine, the electron acceptor, is the **oxidising agent** because it oxidises the sodium. Because an electron transfer involves both a donor and an acceptor, oxidation and reduction must always be linked. Not all redox reactions involve the complete transfer of electrons from one molecule to another, but rather can involve a change in the degree of electron sharing in covalent bonds.

TAKING IT FURTHER:
Further oxidation (p. 156)

10.2 Half-reactions

Since all redox reactions involve both an oxidation and a reduction, we can describe such reactions by two **half-equations** or **half-reactions**. In the reaction above representing the formation of sodium chloride, the sodium atom has been oxidised and the chlorine molecule reduced. We can write half-equations to represent these two reactions:

$Na \rightarrow Na^+ + e^-$ (i) oxidation of sodium

$\frac{1}{2}Cl_2 + e^- \rightarrow Cl^-$ (ii) reduction of chlorine

If we add together the right-hand sides and the left-hand sides of equations (i) and (ii) the electrons cancel out and we obtain the overall equation for the reaction.

$Na + \frac{1}{2}Cl_2 \rightarrow Na^+ Cl^-$ (i) + (ii) overall equation

Consider a biological redox couple – the cellular conversion of malate to oxaloacetate. The reaction is catalysed by the enzyme malate dehydrogenase and involves the coenzyme NAD^+ (nicotinamide adenine dinucleotide). The reaction can be written as:

$$\text{malate} + NAD^+ \longrightarrow \text{oxaloacetate} + NADH + H^+$$

Considering the structures of malate and oxaloacetate (Fig. 82), malate has lost two protons and *two electrons* overall and so has been oxidised in the conversion to oxaloacetate. The reaction actually involves the removal of a hydride and a proton from malate; a hydride (H^-) is a proton with two electrons ($H^+ + 2e^-$).

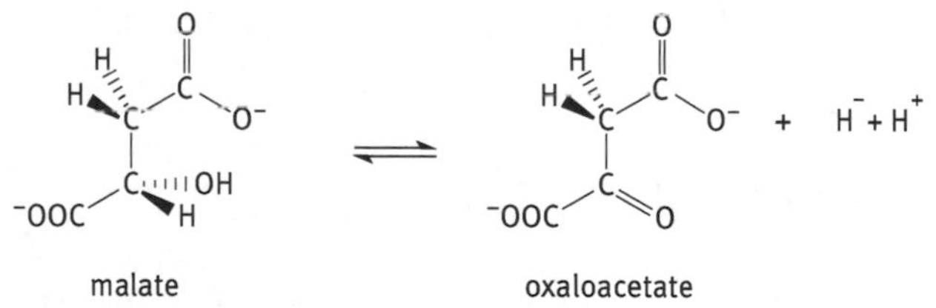

malate ⇌ oxaloacetate + $H^- + H^+$

Figure 82. Oxidation of malate to oxaloacetate

QUESTION

Why has malate lost two electrons?

A proton H^+ has been removed from the hydroxyl group, and a hydride H^- has been removed from the -C-H. The 'H^-' has taken both electrons from the C-H bond with it

In the second half-reaction (Fig. 83), NAD^+ is reduced to NADH. NAD^+ *accepts two electrons* (from the hydride, H^-) and one proton and so is clearly reduced.

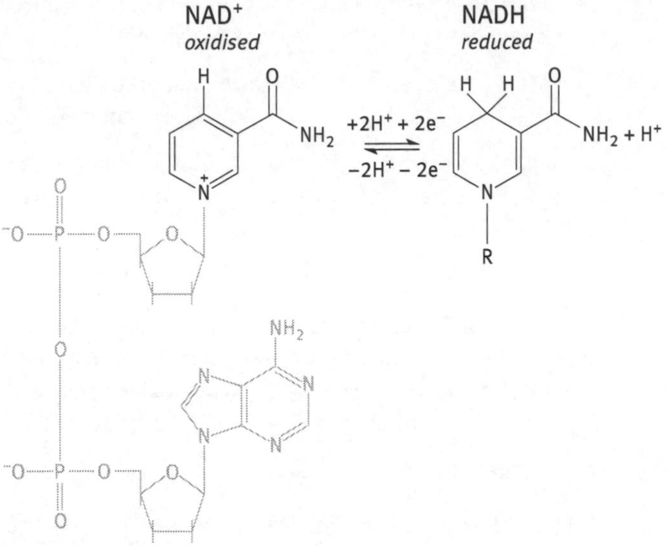

Figure 83. Reduction of NAD^+ to NADH

The two half-reactions are clearly linked, and together they constitute a **redox couple**; we can add them together to get an overall equation. The malate–oxaloacetate reaction is reversible, as indeed are most redox reactions.

10.3 Redox potential

As we have previously seen, the thermodynamic potential of a chemical reaction is calculated from a knowledge of both its equilibrium constant, derived from the concentrations of reactants and products, and from the free energy change (ΔG) of the reaction. All redox reactions on the other hand involve the loss and gain of electrons. It is clearly impractical to measure electron concentrations directly. Instead, we measure the **redox potential** of each half-reaction. The redox potential is a measure of the tendency for a species to either gain or lose electrons.

The redox potential of a half-reaction must be measured relative to a reference; that reference, a standard half-reaction, is the hydrogen electrode, which is arbitrarily set at 0 volts. Redox potentials are measured in volts and chemists designate this by $\boldsymbol{E}^{\circ}$; in biological systems we measure redox potentials at physiological pH and use the symbol $\boldsymbol{E}^{\circ\prime}$ to denote this difference. Furthermore, at pH 7, the redox potential of the hydrogen electrode converts to –0.42 V; this is the value biologists use as their standard redox potential ($\boldsymbol{E}^{\circ\prime}$). A measured redox potential which has a negative value for $\boldsymbol{E}^{\circ\prime}$ indicates the tendency of the reaction to proceed in the direction of oxidation, i.e. to lose electrons. A positive value for $E^{\circ\prime}$ suggests that the reaction is likely to be one of reduction, i.e. a gain in electrons.

In the table below, redox potentials are given for a number of common biological half-reactions. Notice that all of these are shown as reductions. This does not imply that this is the preferred direction of each reaction, but is simply a convention that chemists use in reporting such data.

Redox potentials of some common biological half-reactions

Redox half-reaction	$E^{\circ\prime}$/volt	
$2H^+ + 2e^- \longrightarrow H_2$	−0.42	low
ferredoxin(Fe^{3+}) + e^- ⟶ ferredoxin(Fe^{2+})	−0.42	redox
$NAD^+ + 2H^+ + 2e^- \longrightarrow NADH + H^+$	−0.32	potential
$S + 2H^+ + 2e^- \longrightarrow H_2S$	−0.274	↑
$SO_4^{2-} + 10H^+ + 8e^- \longrightarrow H_2S + 4H_2O$	−0.22	
acetaldehyde + $2H^+ + 2e^-$ ⟶ ethanol	−0.20	
pyruvate + $2H^+ + 2e^-$ ⟶ lactate	−0.185	
$FAD + 2H^+ + 2e^- \longrightarrow FADH_2$	−0.18	
oxaloacetate + $2H^+ + 2e^-$ ⟶ malate	−0.17	
fumarate + $2H^+ + 2e^-$ ⟶ succinate	0.03	
cytochrome *b*(Fe^{3+}) + e^- ⟶ cytochrome *b*(Fe^{2+})	0.075	
ubiquinone + $2H^+ + 2e^-$ ⟶ ubiquinone H_2	0.10	
cytochrome *c*(Fe^{3+}) + e^- ⟶ cytochrome *c*(Fe^{2+})	0.254	
$NO_3^- + 2H^+ + 2e^- \longrightarrow NO_2^- + H_2O$	0.421	↓
$NO_2^- + 8H^+ + 6e^- \longrightarrow NH_4^+ + 2H_2O$	0.44	high
$Fe^{3+} + e^- \longrightarrow Fe^{2+}$	0.771	redox
$O_2 + 4H^+ + 4e^- \longrightarrow 2H_2O$	0.815	potential

A reaction which we have used before, namely the reduction of pyruvate to lactate, is an example of a redox reaction.

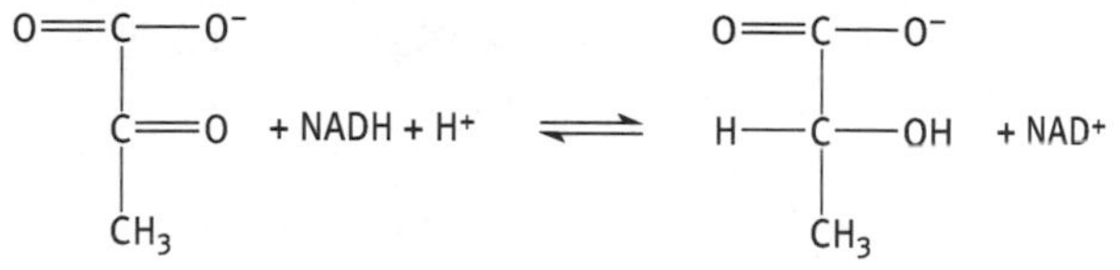

$$\underset{\text{pyruvate}}{CH_3{-}CO{-}COO^-} + NADH + H^+ \rightleftharpoons \underset{\text{lactate}}{CH_3{-}CH(OH){-}COO^-} + NAD^+$$

The two half-reactions of this redox couple are also given in the table. That for pyruvate reduction is:

$$\text{pyruvate} + 2H^+ + 2e^- \longrightarrow \text{lactate} \qquad (1)$$

The redox potential for this half-reaction is −0.185 V (read from the table).

For NADH the half-reaction, which is an oxidation reaction, is written as:

$$NADH + H^+ \longrightarrow NAD^+ + 2H^+ + 2e^- \qquad (2)$$

The half-reaction for NAD^+/NADH shown in the table is a reduction reaction (by convention). To obtain the redox potential for the oxidation of NADH, we simply reverse the sign of the $E^{\circ\prime}$ given in the table. Therefore, the redox potential for the oxidation of NADH is +0.32 V (rather than −0.32 V). When we add the redox potentials for both half-reactions together, to find the overall redox potential of the enzyme-catalysed reaction, we get:

$$-0.185\ \text{V} + 0.32\ \text{V} = +0.135\ \text{V}$$

When we add the right-hand sides and the left-hand sides of equations (1) and (2) the electrons cancel and we get:

$$\text{pyruvate} + \text{NADH} + \text{H}^+ \rightleftharpoons \text{lactate} + \text{NAD}^+ \quad E^{\circ\prime} = 0.135\ \text{V}$$

10.4 Free energy and redox potentials

As mentioned earlier, it is impractical to measure an electron concentration in order to derive an equilibrium constant for a redox reaction. However, by derivation we find that the free energy of a redox reaction can be calculated directly from its $E^{\circ\prime}$ by the **Nernst equation.** The Nernst equation is shown as:

$$\Delta G^{\circ\prime} = -nF\,\Delta E^{\circ\prime}$$

where n is the number of electrons transferred in the reaction and F is the Faraday constant (23.06 kcal V^{-1} mol^{-1} or 96.5 kJ V^{-1} mol^{-1}).

Using determinations of equilibrium constants (K_{eq}) and redox potentials ($E^{\circ\prime}$), biologists can therefore determine the $\Delta G^{\circ\prime}$ for a variety of reactions with a view to assessing their feasibility and directionality.

10.5 Obtaining energy for life

Biological molecules which are capable of accepting and donating electrons, being alternately reduced and oxidised, are referred to as **electron carriers.** NADH is an example of an electron carrier. Dependent on their redox potentials, we could arrange a series of different electron carriers to form an **electron transport chain.** Electrons could pass down such a chain, passing from electron carriers with a low redox potential (more negative and therefore more likely to become oxidised by losing an electron) to ones with an increasingly high redox potential (more positive). Electron carriers with a low (more negative) redox potential have a lower affinity for the electron and will pass it on to a carrier with a higher (more positive) redox potential and greater affinity for the electron.

In Section 8.7 we indicated that the reduction of oxygen (or the oxidation of hydrogen) in the reaction

$$2H_2 + O_2 \rightarrow 2H_2O$$

was highly exergonic and spontaneous, liberating some 447 kJ mol^{-1} of energy, mostly as heat and sound. The subcellular organelle, the mitochondrion, carries out an analogous process whereby a series of electron carriers provide a carefully controlled pathway in which free energy can be extracted in small packages. At one end of the **'electron transport chain'** NADH , which is generated in the cell by catabolic pathways, is oxidised and the two electrons 'lost' in this oxidation are passed down the chain, from one electron carrier to the next, finally ending in a reaction which results in the reduction of oxygen to water (which is why you need to breathe oxygen to survive!).

The oxidation of NADH, with a relatively low redox potential, and the reduction of oxygen, with a relatively high redox potential, creates a potential difference from 'top to bottom' of the chain of some +1.13 V

$$\tfrac{1}{2}O_2 + 2H^+ + 2e^- \rightarrow H_2O \qquad E^{\circ\prime} = +0.815\ \text{V}$$

$$\text{NADH} + H^+ \rightarrow \text{NAD}^+ + 2H^+ + 2e^- \qquad E^{\circ\prime} = +0.32\ \text{V}$$

The overall reaction is:

$$\tfrac{1}{2}O_2 + \text{NADH} + H^+ \rightarrow H_2O + \text{NAD}^+ \qquad \Delta E^{\circ\prime} = +1.13\ \text{V}$$

Inserting this value of $E^{\circ\prime}$ into the Nernst equation gives a free energy change of −218 kJ mol^{-1} .

$$\Delta G^{\circ\prime} = -nF\,\Delta E^{\circ\prime} = -2(96.5\ \text{kJ V}^{-1}\ \text{mol}^{-1})\ (1.13\ \text{V}) = -218\ \text{kJ mol}^{-1}$$

Note that n, the number of electrons involved in the reaction, is 2 in this case.

Not only is this series of reactions highly exergonic, releasing a considerable amount of free energy, but the passage of electrons 'down' this chain is driven by a significant potential difference of +1.13 V.

A redox couple such as these two half-equations, which together comprise the biological reaction between oxygen and hydrogen to produce water, form a link in an electron transport chain. Such a link can be represented by the following convention.

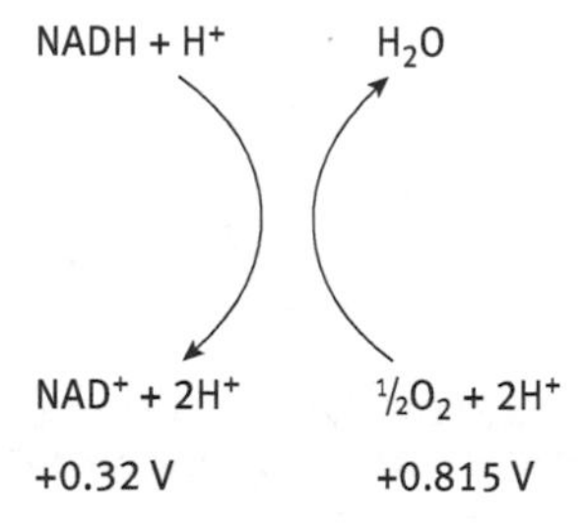

The curved arrows represent the movement of electrons down the chain as each species is oxidised.

10.6 What happens to this free energy?

The mitochondrial electron transport chain is depicted in Fig. 84; the various electron carriers are shown in a linear fashion. Each redox couple is associated with a redox potential ($\Delta E^{\circ\prime}$) and a calculated $\Delta G^{\circ\prime}$. As we pass along the chain from left to right, the redox potential becomes progressively more positive (with a greater tendency to become reduced), whereas we note that there are three redox couples in particular which are associated with much higher free energy changes; these are the oxidation of NADH and reduction of FAD, oxidation of cytochrome *b* and reduction of cytochrome *c*, and the oxidation of cytochrome *a* and reduction of oxygen. The large negative free energy change at these three sites is used to do work; ultimately this results in the synthesis of ATP. ATP is the cells' energy currency; the free energy stored in this molecule can be coupled to unfavourable endergonic reactions to drive anabolic processes.

TAKING IT FURTHER:
Free energy and metabolic pathways (p. 125)

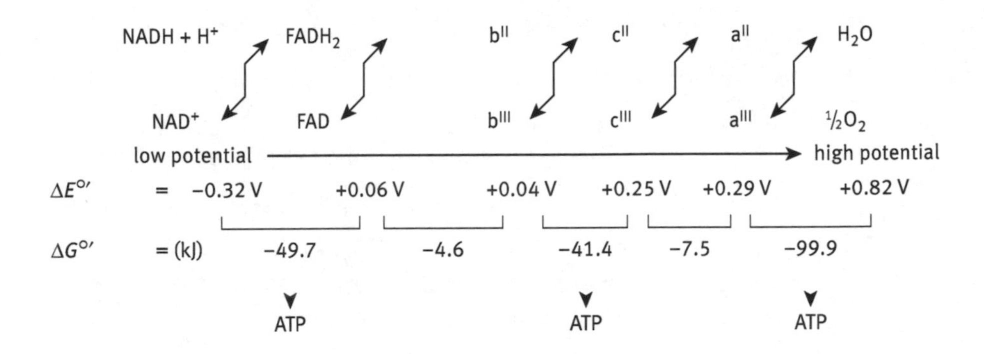

Figure 84. Redox and free energy steps in the mitochondrial electron transport chain

[Key to the mitochondrial electron transport chain: NAD = nicotinamide adenine dinucleotide; FAD = flavin adenine dinucleotide; b = cytochrome *b*; c = cytochrome *c*; a = cytochrome *a*; b^{II}, c^{II} and a^{II} represent the reduced cytochrome, and b^{III}, c^{III} and a^{III} the oxidised cytochrome]

Mitochondria are the principal site for ATP generation in higher organisms. This aerobic process, referred to as oxidative phosphorylation, is essential for energy provision in all higher organisms.

REMINDER

The oxidative phosphorylation process is a series of oxidation steps which eventually bring about the synthesis of ATP through the phosphorylation of ADP

10.7 Summing up

1. Oxidation always involves the removal of electrons from a molecule. Whenever one molecule is oxidised another must be reduced. Thus we speak of redox reactions.
2. All redox reactions are made up from two half-reactions, one half-reaction involving an oxidation and the other involving a reduction. For each half-reaction we may measure a redox potential; the redox potential is a measure of the degree to which a reaction will either gain or lose electrons. It is measured relative to a reference; that reference, a standard half-reaction, is the hydrogen electrode, which is arbitrarily set to 0 volts. Redox potentials are therefore measured in volts and are designated by E°.
3. A negative value for $E^{\circ\prime}$ indicates the tendency of the reaction to proceed in the direction of oxidation, i.e. to lose electrons.
4. The free energy change of a redox reaction is related to the redox potential by the Nernst equation, $\Delta G^{\circ\prime} = -nF\Delta E^{\circ\prime}$
5. Electrons will transfer from a low redox potential (more negative $E^{\circ\prime}$) to a higher redox potential (more positive $E^{\circ\prime}$). This transfer of electrons is associated with a free energy change.
6. The subcellular organelle, the mitochondrion, uses an arrangement of electron carriers to mediate the transfer of electrons from a low to a higher redox potential, and utilises the free energy released to do work and eventually generate ATP in the process of oxidative phosphorylation.

10.8 Test yourself

The answers are given on page 182.

Question 10.1
Write half-reactions for the following redox reactions
(a) $Zn + Cu^{2+} \rightarrow Zn^{2+} + Cu$
(b) $Fe^{2+} + Cu^{2+} \rightarrow Fe^{3+} + Cu^{+}$

Question 10.2
The following reaction shows the reduction of acetaldehyde to form ethanol, catalysed by the enzyme alcohol dehydrogenase.
acetaldehyde + NADH + $H^{+} \rightleftharpoons$ ethanol + NAD^{+}
(a) Write the two half-reactions for this redox couple.
(b) Using the redox potentials tabulated on page 151, decide whether the above reaction will proceed in a forward (left to right) or backward direction.

Question 10.3
Using the redox potentials tabulated on page 151, decide in which direction the following redox couple will proceed.
ubiquinone $_{(oxidised)}$ + cytochrome *c* $_{(reduced)}$ $\rightleftharpoons$ ubiquinone $_{(reduced)}$ + cytochrome *c* $_{(oxidised)}$

Question 10.4
Using the equation $\Delta G^{\circ\prime} = -nF\Delta E^{\circ\prime}$, calculate the free energy change for the reaction:
succinate + FAD $\rightleftharpoons$ fumarate + $FADH_2$
(use tabulated values on page 151 for the redox potentials; number of electrons transferred in this reaction = 2; take F as 96485 J V^{-1} mol^{-1})

Question 10.5
Using the equation $\Delta G^{\circ\prime} = -nF\Delta E^{\circ\prime}$, calculate $\Delta G^{\circ\prime}$ for the following reaction:
malate + $NAD^{+} \rightleftharpoons$ oxaloacetate + NADH

Taking it further

Further oxidation

Just as electrons in atoms occupy atomic orbitals, such that the lowest energy orbitals are filled first, so in molecules electrons occupy molecular orbitals. Molecular orbitals are regions of space in which there is a certain probability of finding an electron. Electrons in molecules possess energy and fill molecular orbitals such that a minimum energy state is adopted. The higher in energy the molecular orbital, the more energy the electron possesses.

We have defined **oxidation** as the loss of electrons. Stated slightly differently, oxidation is the movement of electrons *away from* an atom (or, more precisely, away from an atom's nucleus). Thus, when oxygen oxidises something, it pulls the electrons away from that something and towards itself (in the process oxygen serves as an oxidising agent and is itself reduced). Remember, oxygen is an electronegative atom. Note that oxidation refers to this movement of electrons even when oxygen atoms are not involved in the process.

In Sections 2.5 and 2.6 we considered **polar covalent bonds** which arise when a more **electronegative** atom, such as oxygen, is bonded to a less electronegative atom such as hydrogen or carbon. The more electronegative atom draws electrons towards it (and is reduced) whilst the other atom in the bond, which 'loses' electrons, is oxidised.

Consider the succession of compounds in Fig. 85.

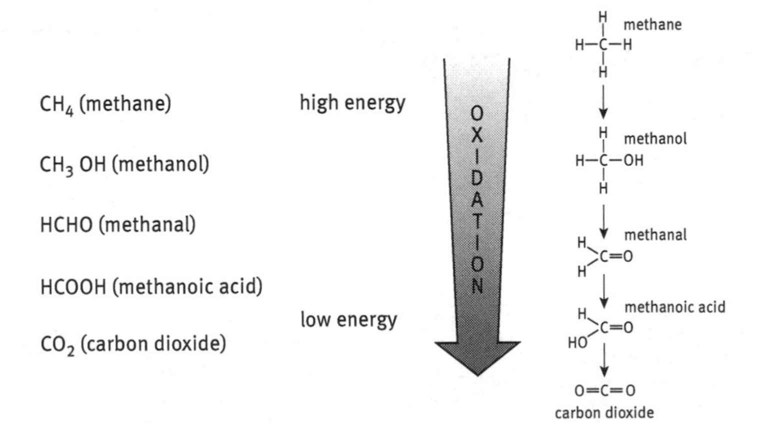

Figure 85. Oxidation of high energy compounds

In carbon–hydrogen bonds (e.g. methane), the electronegativity of carbon and hydrogen are approximately equal. Therefore the electrons in this symmetrical covalent bond are as far from the nuclei of the two atoms as is possible. They are more or less shared evenly between carbon and hydrogen. As we pass down this series, hydrogen atoms are progressively replaced by electronegative oxygen, and the carbon atom is progressively oxidised by the formation of polar covalent bonds (in which electrons are drawn away from the carbon towards the oxygen). It must follow therefore that the electrons in the methane covalent bonds possess more energy than those in carbon dioxide (since through oxidation they are found progressively closer to the oxygen atomic nucleus). Energy is stored in the carbon–hydrogen bonds of methane and methane is a **high energy** compound, whereas carbon dioxide is a **low energy** compound.

In the process of photosynthesis, plants capture light energy and incorporate this primarily into carbohydrates. Carbohydrates produced by photosynthesis, such as glucose, mainly consist of carbon–carbon and carbon–hydrogen covalent bonds. We might consider these symmetrical covalent bonds as **'energy-rich bonds'**, in so much as the electrons in them possess a high energy. Captured solar energy is stored in carbon compounds, such as cellulose in trees; over time these have become buried and now provide us with our hydrocarbon reserves of oil, gas and coal so-called 'fossil fuels'. **Hydrocarbons** are so called because they consist of only carbon and hydrogen. Some 'energy-rich' compounds are shown in Fig. 86.

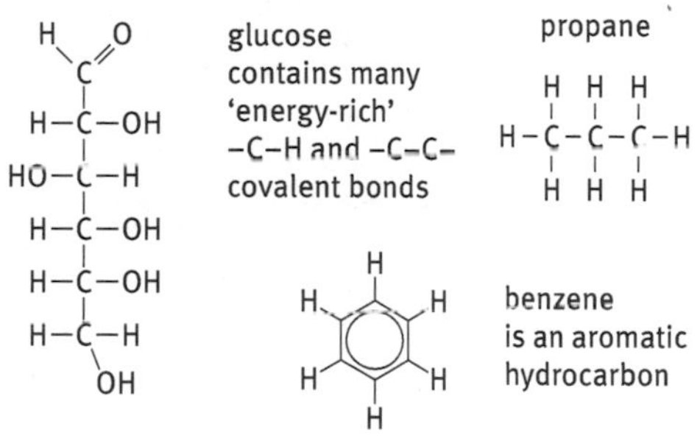

glucose contains many 'energy-rich' –C–H and –C–C– covalent bonds

propane

benzene is an aromatic hydrocarbon

Figure 86. High-energy glucose and hydrocarbons

The first law of thermodynamics states that energy can neither be created nor destroyed. The transfer of an electron from a less electronegative atom to a more electronegative atom must therefore result in some transfer of energy from that electron to the surrounding environment. When we burn fossil fuels in air, we oxidise them and the energy is released as heat. When the mitochondria in our cells 'burn' fuels they likewise oxidise them, but the energy is released step-wise, harnessed and stored in ATP for future use. It is worth pointing out that this mitochondrial transfer of energy to ATP is far from being 100% efficient, indeed much is lost as heat. As warm-blooded animals we owe most of our heat to this oxidation process.

11 Reactivity of biological molecules

BASIC CONCEPTS:

The presence of functional groups on biomolecules provides reactive sites, at which such molecules may be transformed or linked to other molecules. Reactive sites may be nucleophilic or electrophilic, and the reactions which occur at such sites can be classified as addition, substitution or elimination reactions. Functional groups are essential in conferring a 'site of attack' for enzymes; reactions at a nucleophilic or electrophilic centre often generate intermediates or transition states which can be stabilised by an enzyme. It is by these means that enzymes provide alternative and feasible reaction routes which underline their role as catalysts.

Understanding reaction mechanisms allows biologists and chemists to optimise the yield of a chemical in a biotechnological industrial process, or understand catalysis by metals and by enzymes, and to design inhibitors of enzymes, such as drugs and pesticides.

The metabolic reactions which occur in organisms consist of many apparently very complex transformations. Biomolecules consist of a large number of atoms and bonds at any one of which a chemical reaction could occur. Fortunately, however, we know that the **reactive sites** on such molecules invariably involve their **functional groups**. The functional group behaves differently from the rest of the molecule, usually because it contains an electronegative atom and consequently a polar covalent bond. As we have previously discussed in Chapter 3, functional groups are vitally important in the provision of intermolecular interactions which stabilise molecular interactions and conformations (shapes) of molecules. But in providing reactive sites, functional groups also provide the means for molecular transformations, as well as the means for molecular building and construction of biological **macromolecules**.

Reactive sites on a molecule may contain a **nucleophilic centre** or an **electrophilic centre**. Nucleophilic centres are electron rich; they may have a negative charge, have lone pairs of electrons, or possess an increased electron density typical of double covalent bonds. A nucleophilic centre will attract positively charged groups, i.e. an electrophile (for example, a positively charged proton). Electrophilic centres are electron deficient and seek out negative charges.

Take for example the carbonyl group. Carbonyl groups (C=O) are found in many biological molecules containing:

- RCOOH (carboxylic acid), in acids
- RCHO (aldehyde), in sugars
- R_2CO (ketone), in sugars
- $RCONH_2$ (amide), in proteins

The C=O group is polar, with a small positive charge (δ+) on C and a small negative charge (δ–) on O. Nucleophiles are attracted to the slightly positive C, forcing electrons upon it. Electrophilic protons are attracted to the slightly negative O, which can donate electrons.

REMINDER

An electronegative atom is frequently found as part of a functional group, which generates a charge dipole in that group, and subsequently results in both a nucleophilic and an electrophilic centre.

Nucleophilic and electrophilic reactive sites provide the basis of a number of different types of reaction mechanisms.

There is a specific convention for showing the route or mechanism by which a chemical reaction takes place. Bond breaking and bond making involve the transfer of electrons and the convention described here, which uses curly arrows, allows us to keep track of the electrons.

$$X{:}Y \longrightarrow X^{\oplus} + {:}Y^{\ominus}$$

The bond breaking step shown above involves both the electrons from the covalent bond X–Y being transferred to the atom Y. This type of bond breaking is known as **heterolytic** bond breaking because both electrons migrate to the same atom. The position and nature of the curly arrow is important. The double-headed curly arrow represents the fact that both electrons in the bond are moving. The arrow starts at the pair of electrons in the bond and finishes at the atom Y which has accepted the electrons. Complete transfer of two electrons usually generates ions. In the mechanisms shown, the pairs of electrons are not always shown but the double-headed arrow implies the movement of two electrons, as bonds are either formed or broken.

In some cases chemical bonds are broken so that one electron transfers to each of the atoms at either end of the bond. This type of bond-breaking is called **homolytic** bond cleavage and half-headed arrows are used to represent the movement of electrons. The species formed are called **free radicals** and are chemically very reactive.

$$X:Y \longrightarrow X^{\bullet} + Y^{\bullet}$$

11.1 Addition reactions

The reaction sequence in Fig. 87 shows a ketone called propanone (acetone) forming a hydrate under acidic conditions. The nucleophilic OH^- attacks the electrophilic carbon of the ketone group, whilst an electrophilic proton attacks the nucleophilic oxygen of the ketone group. The large curved arrow in this diagram indicates the first point of attack by the nucleophile; the short curly arrow is used to indicate the movement of a pair of electrons, in this case two electrons from the pi covalent bond to form a new sigma covalent bond with the proton.

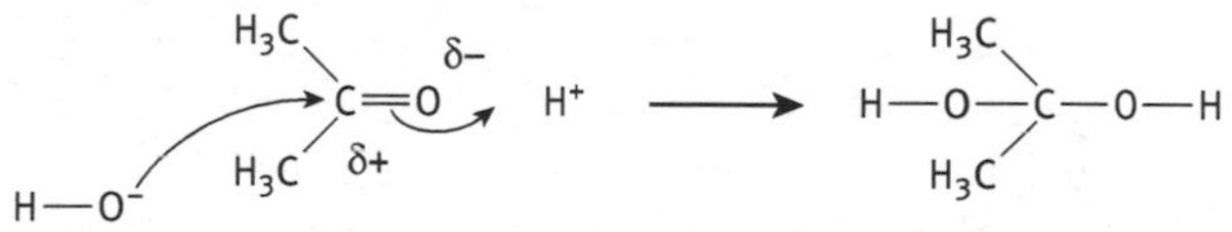

$$H{-}O^- + (H_3C)_2C^{\delta+}{=}O^{\delta-} + H^+ \longrightarrow H{-}O{-}C(CH_3)_2{-}O{-}H$$

Figure 87. An addition reaction to a ketone

This type of reaction is referred to as an **addition reaction** (OH^- and H^+ have both been added to the ketone).

When reactions involve either nucleophiles or electrophiles, then the electrons involved in forming new bonds or breaking existing bonds move in pairs.

11.2 Substitution reactions

Shown in Fig. 88 is the mechanism of the reaction of bromoethane in alkaline solution to produce ethanol. In bromoethane the carbon is an electrophilic centre, because of the polar covalent bond with bromine (bromine is a more electronegative atom). A nucleophilic hydroxide ion is attracted to the electrophilic carbon centre, causing the electron pair of the C-Br polar covalent bond to move onto the bromine, resulting in a substitution of Br by OH and formation of the alcohol and a Br^- ion.

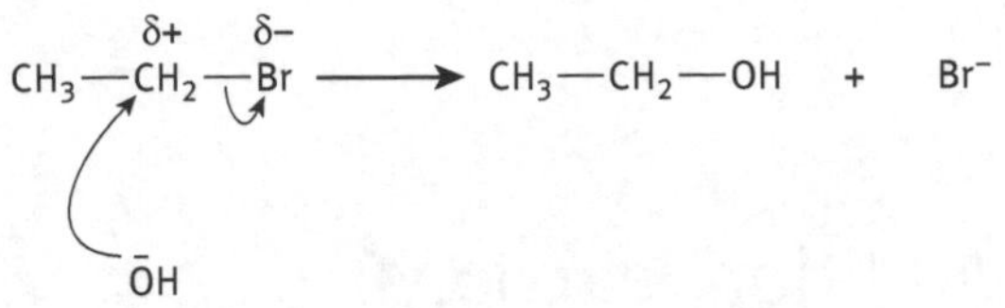

$$CH_3-\overset{\delta+}{CH_2}-\overset{\delta-}{Br} \longrightarrow CH_3-CH_2-OH + Br^-$$

$$\bar{O}H$$

Figure 88. Formation of an alcohol by a substitution reaction

This is an example of a **nucleophilic substitution.**

11.3 Elimination reactions

Consider the dehydration (removal of water) from ethanol in acidic solution to produce ethene (Fig. 89).

The lone pair of electrons on the oxygen atom moves to form a coordinate covalent bond with an H^+ ion (both electrons in the bond are provided by the oxygen). This momentarily imparts a positive charge on the oxygen, causing an electron pair to move from the C-O bond to reinstate the lone pair of electrons on the oxygen, in the process resulting in the formation and loss of water from the ethanol molecule, but also forming an electrophilic centre on the carbon atom. This unstable intermediate molecule rearranges further, with an electron pair moving from a C-H bond of the methyl (CH_3) group to the C^+, so satisfying the tetravalency of carbon and resulting in the elimination of H^+ and formation of ethene.

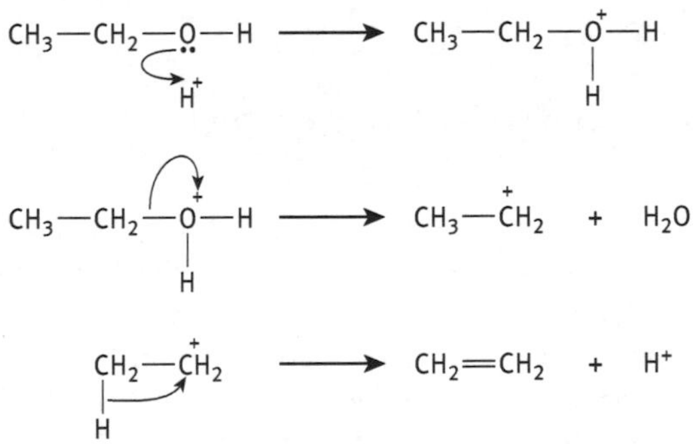

$$CH_3-CH_2-\ddot{O}-H + H^+ \longrightarrow CH_3-CH_2-\overset{+}{O}(H)-H$$

$$CH_3-CH_2-\overset{+}{O}(H)-H \longrightarrow CH_3-\overset{+}{C}H_2 + H_2O$$

$$H-CH_2-\overset{+}{C}H_2 \longrightarrow CH_2{=}CH_2 + H^+$$

Figure 89. The dehydration of ethanol by an elimination reaction

REMINDER

In addition, substitution and elimination reactions, the electrons involved in forming new bonds or breaking existing bonds always move in pairs

11.4 Reaction mechanisms and reaction kinetics

We saw in Chapter 9 that the rate at which a chemical reaction takes place can be represented by the rate equation for the reaction. The rate equation relates to the slowest step in the mechanism for the chemical reaction. For a general reaction

$$A + B \rightarrow C$$

the rate equation may be given by:

$$\text{Rate} = k[A][B]$$

This implies that the reaction is second order overall and the rate of reaction depends upon the concentrations of both A and B.

Alternatively the reaction may take place by a different mechanism in which the slow step depends only upon the concentration of one of the reactants. In this case the kinetics would be first order and the rate equation given by:

$$\text{Rate} = k[A] \text{ or Rate} = k[B]$$

Substitution and elimination reactions can proceed by different mechanisms which are either first or second order.

The mechanism shown in Fig. 88, in which ethanol is formed by a substitution reaction in which the hydroxide ion displaces a bromide ion from bromoethane, is second order. The step shown involves the attack of the nucleophile, OH^-, on the bromoethane and is therefore second order. Because this is a second order nucleophilic substitution reaction it is described as an S_N2 reaction.

There is an alternative mechanism for nucleophilic substitution reactions of this type, known as S_N1 reactions (Fig. 90). Such reactions are first order and the kinetics depend only on the concentration of the halogeneoalkane. The substitution of bromine by OH^- in tertiary butyl bromide to give tertiary butanol proceeds in this way.

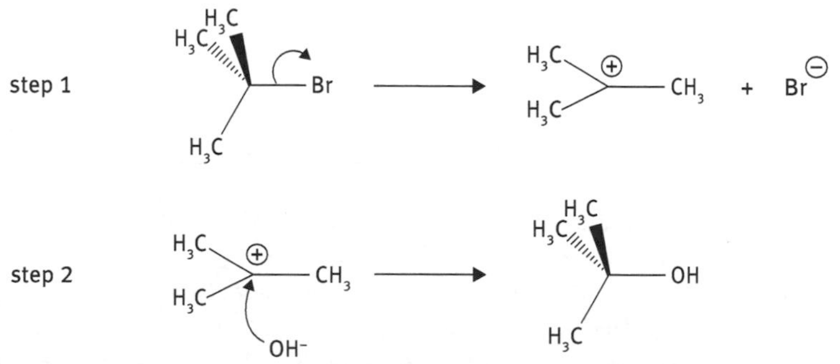

step 1

step 2

Figure 90. Formation of t-butanol by a substitution reaction (S_N1 reaction)

In step 1 the bromine atom is lost as the C-Br bond dissociates leaving a planar carbenium ion. This is a unimolecular reaction and is slow. Once formed, the carbenium ion is reactive and rapidly attacked by nucleophiles such as the hydroxide ion or water. Overall replacement of bromine by the hydroxyl group to give a tertiary alcohol has occurred.

Reactions involving molecules with bulky substituents tend to occur via the S_N1 mechanism because of steric hindrance as incoming nucleophiles are physically prevented from attacking the electropositive carbon atom. However, there are a variety of other factors which determine the reaction mechanism.

11.5 Free radical reactions

Normally, bonds do not split in a way that leaves an atom or molecule with an odd, unpaired electron. However, **free radicals** do contain a single unpaired electron, and are produced by the **homolytic** fission of a covalent bond. Free radicals are indicated by a single dot to the side and middle of the atom concerned; for example, the chlorine free radical is shown as **Cl·**

In Fig. 91, the first reaction shown is a **heterolytic** fission, the 'full head' curly arrow denoting the movement of a **pair** of electrons; the products formed are **ions**. The second reaction is a homolytic reaction, the 'half head' curly arrows denoting the movements of single electrons; the products formed are free radicals.

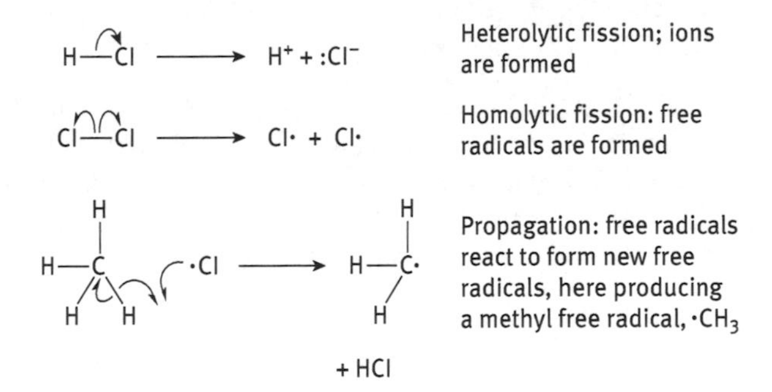

Figure 91. Heterolytic and homolytic fission of covalent bonds

Free radicals are very unstable and react quickly with other compounds, trying to capture the needed electron to gain stability. Generally, free radicals attack the nearest stable molecule, 'stealing' an electron. When the 'attacked' molecule loses an electron, it becomes a free radical itself, beginning a chain reaction and **propagating** the reaction to other molecules. Once the process is started, it can cascade resulting in damage and disruption to the living cell.

Some free radicals arise normally during metabolism. Indeed, the body's immune system cells purposefully create them to neutralise viruses and bacteria. However, environmental factors such as pollution, radiation, cigarette smoke and herbicides can also spawn free radicals. **Antioxidants** include a range of different chemicals which can **terminate** free radical propagation; they do this by donating an electron, so ending the 'electron stealing' reaction chain. Vitamins C and E are examples of biological antioxidants.

REMINDER

Free radicals are generated by the homolytic cleavage of a covalent bond and contain a single unpaired electron

11.6 Pi bonds and addition reactions

Perhaps not obviously so, the pi bond may also be considered to be a functional group, across which addition reactions can occur. Remember that a covalent double bond consists of a sigma and a pi bond. The pi bond is 'electron rich' and can effectively act as a nucleophilic site (Fig. 92). The pi bond often breaks and the electrons in it are used to join other atoms or groups in an addition reaction.

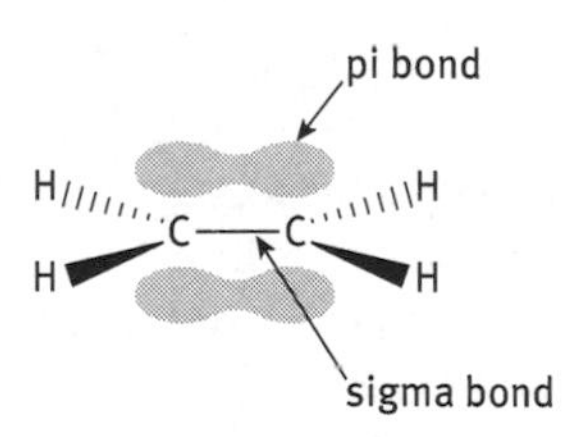

Figure 92. The 'electron-rich' pi bond is a nucleophilic site

Consider the reaction scheme in Fig. 93 involving a carbon–carbon double bond.

In stage **1**, the pi bond is approached by the molecule X-Y; Y is highly electronegative and so X is an electrophile. As $X^{\delta+}$ approaches the pi bond, electrons in the X-Y bond are further pushed onto Y (shown by ··).

REMINDER

Pi bonds provide an 'electron rich' site which effectively acts as a nucleophilic reactive site

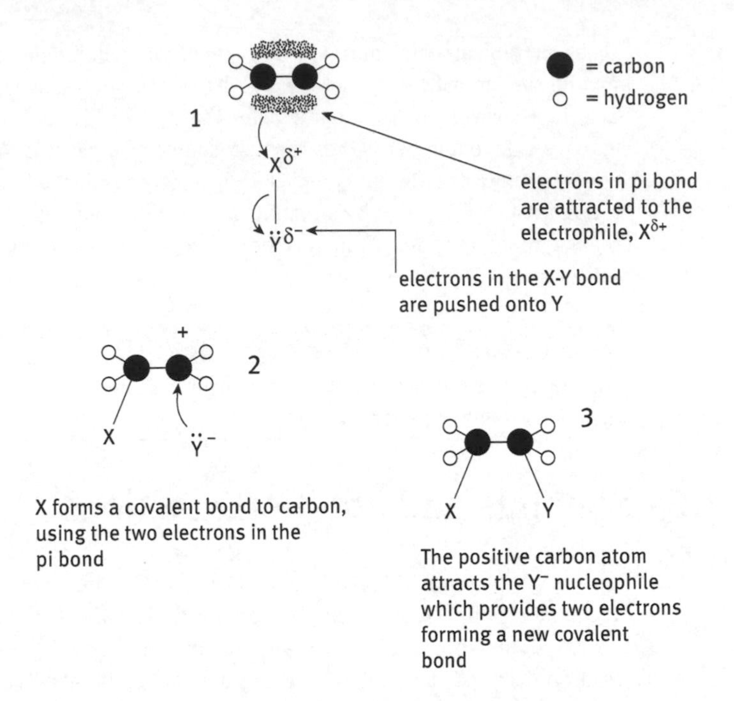

Figure 93. The addition reaction across a pi bond

In stage **2**, X forms a new covalent bond to a carbon atom using the two electrons from the pi bond. The pi bond was originally made using an electron from each carbon atom, but both of these electrons have now been used to make a bond to the X atom. This leaves the right-hand carbon atom an electron short – hence positively charged. In stage **3**, Y has an electron pair, originally used to bond X to Y; this pair of electrons can form a new covalent bond with the 'electron deficient' carbon.

Thus, an addition reaction has been made across the carbon double bond. Pi covalent bonds are a useful 'source' of electrons for addition reactions to occur.

11.7 Functional groups link molecules together

The hallmark of organisms is their ability to link simple molecules together to form **macromolecules.** Carbohydrates, proteins and nucleic acids are **polymers**, large macromolecules formed through the linkage of numerous **monomers.** Macromolecules can be structural (e.g. collagen in bone and connective tissues, cellulose in plants), functional (e.g. enzymes and hormones), deposits of information (e.g. DNA), as well as deposits of energy (e.g. glycogen and starch).

The hydroxyl functional groups on sugar monomers can react through a condensation reaction (dehydration, loss of water) to form sugar linkages, eventually building to macromolecules such as starch. For example amylose, a constituent of starch, consists typically of 200 to 20 000 glucose units (Fig. 94).

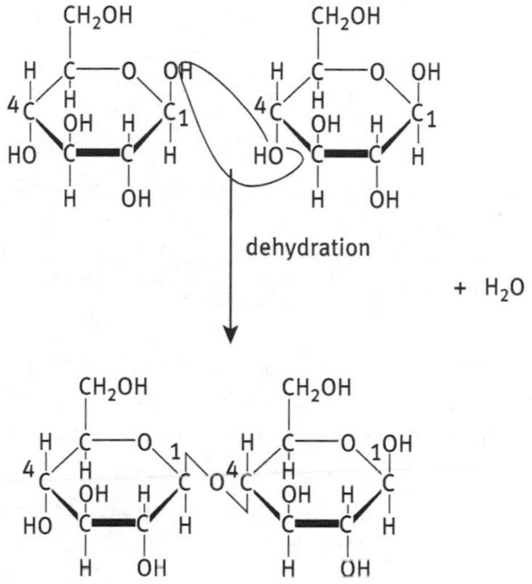

Figure 94. First step in amylose formation through linking of glucose monomers

The bond linkages, between carbon atom 1 of one glucose molecule, and carbon atom 4 of another glucose molecule, are referred to as 1,4-glycosidic bonds.

TAKING IT FURTHER:
The peptide bond
(p. 28)

Similarly, as we have seen previously, amino acids are linked together by peptide bonds through a condensation reaction to form those polymers which are proteins. Again, the reaction involves a functional hydroxyl group, on the 'carboxyl end' of one amino acid, with the elimination of water (Fig. 95).

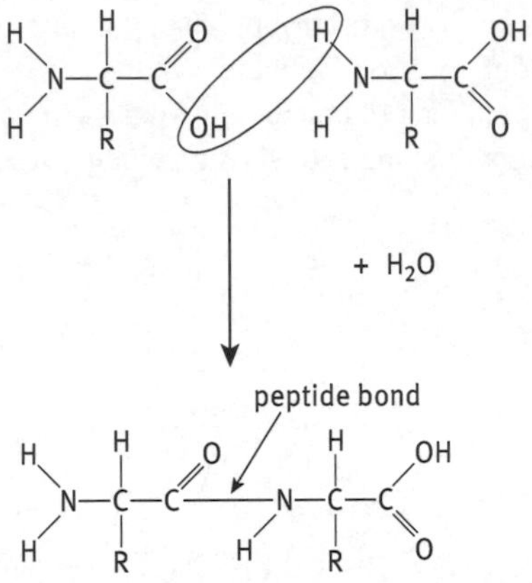

+ H_2O

peptide bond

Figure 95. Amino acid monomers are linked together through peptide bonds

In nucleic acids, the 'backbone' of each single nucleic acid strand is a polymer of nucleotides, linked through dehydration reactions to form phosphodiester bonds between the phosphate group and the 3′ and 5′ carbons of adjacent nucleotides. Once again, functional hydroxyl groups at the 3′ and 5′ positions of the sugar molecule provide the means for such linkages to occur (Fig. 96).

Through these 'polymerisation' reactions, nucleic acids are formed (Fig. 97), the single polymer strands being held together in a double-stranded structure by hydrogen bonding between opposite bases (see Chapter 3).

REMINDER

Carbohydrates, proteins and nucleic acids are polymers, linked together through the functional groups of their monomers

functional group on carbon 5 (links to phosphate)

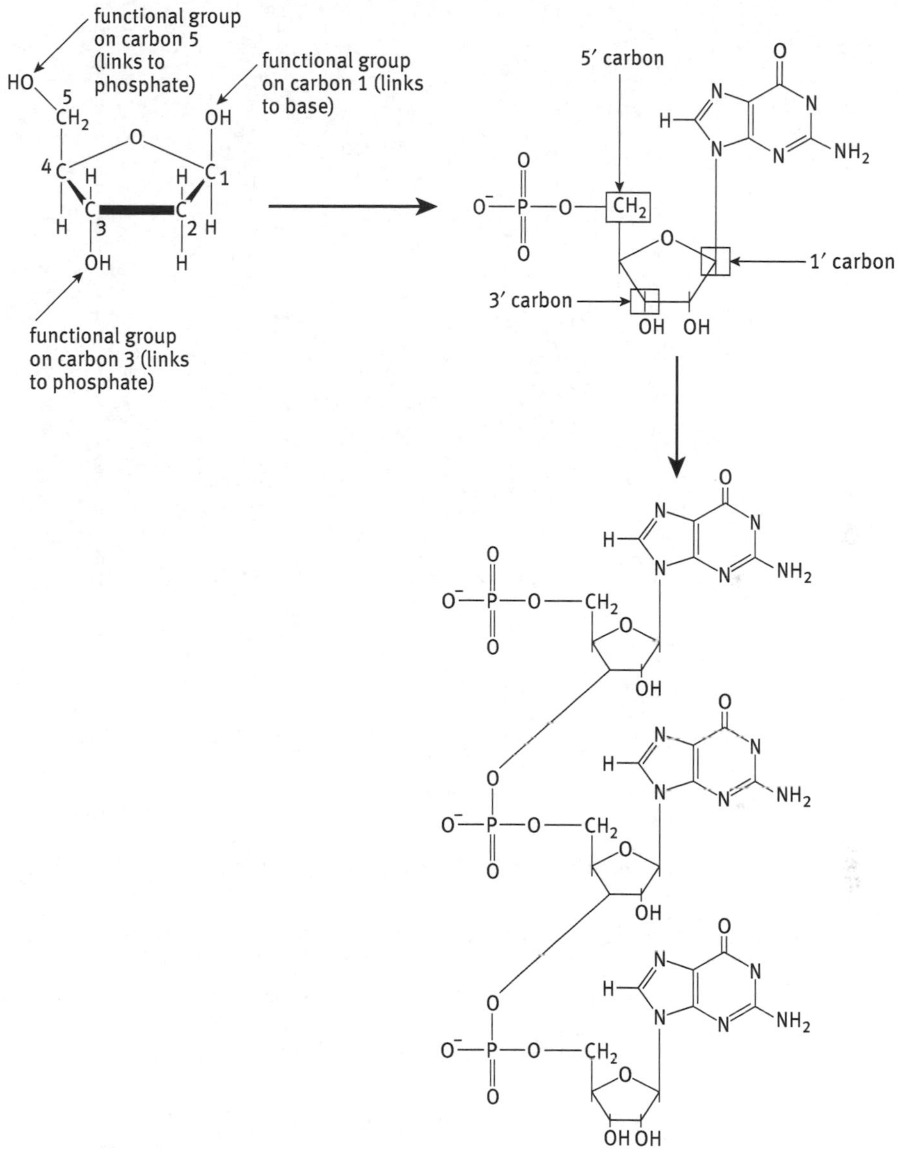

functional group on carbon 1 (links to base)

functional group on carbon 3 (links to phosphate)

5′ carbon

1′ carbon

3′ carbon

Figure 96. Hydroxyl groups on sugar molecules provide the means for constructing nucleotide polymers

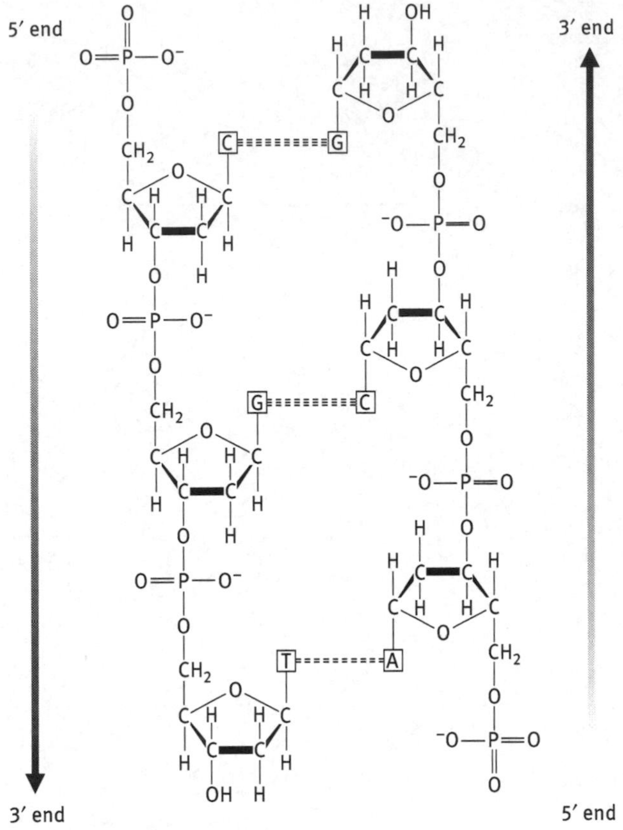

5′ end 3′ end

3′ end 5′ end

Base A = adenine, C = cytosine, G = guanine, T = thymine

Figure 97. Double-stranded nucleic acid

11.8 Enzyme-catalysed reactions

TAKING IT FURTHER: Enzyme catalysis (p. 172)

We considered in Section 8.2 how an enzyme was able to provide an alternative reaction route, and therefore a lower activation energy; it is this ability of enzymes which underlines their roles as biological catalysts. But exactly how is an enzyme able to provide an alternative reaction route? A clue is provided earlier in the elimination reaction scheme for the dehydration of ethanol to ethene (Fig. 89). In this scheme a number of intermediates are shown; such intermediates constitute **transition states** which are generally very unstable and short-lived. Enzymes are able to bind and stabilise transition states; indeed enzymes can make possible different, and often multiple, transition states providing alternative reaction routes from reactant to product with lower activation energies.

11.9 Summing up

1. Functional groups provide reactive sites on biomolecules; these may be nucleophilic or electrophilic centres.
2. Reactions at active sites can be classified as addition, substitution or elimination.
3. Free radicals are produced by the homolytic fission of a covalent bond, as opposed to heterolytic fission which is most common for covalent bond breakage.
4. A pi molecular orbital provides an additional nucleophilic reactive site; the two electrons of the pi bond are used to add additional atoms or groups across the double bond.
5. Functional groups are utilised to join simple molecules (monomers) together to form large macromolecules (polymers). Typical reactions in organisms involve condensation reactions (a dehydration reaction where there is a loss of water), linking monomers through their hydroxyl groups to form macromolecules such as polysaccharides, proteins and nucleic acids.

11.10 Test yourself

The answers are given on pp. 182–183.

Question 11.1
Explain what is meant by a nucleophilic and an electrophilic centre on a molecule.

Question 11.2
Why is a pi bond considered to be a nucleophilic site?

Question 11.3
Explain what is happening in the following sequence.

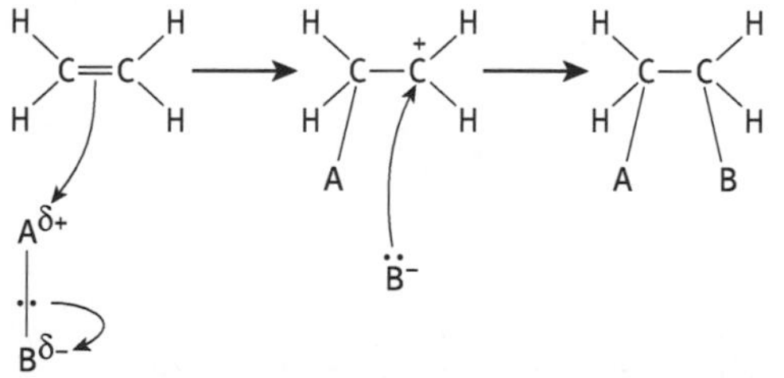

Question 11.4
Explain the relationship between enzymes, activation energy and transition states.

Question 11.5
Explain the difference between an ion and a free radical.

Taking it further

Enzyme catalysis

Enzymes are biological catalysts. They make possible many hundreds of chemical reactions in the cell which might otherwise not be thermodynamically feasible. They are highly specific with regard to which **substrate** they will interact with, and they are controllable making possible the highly complex integration of the cell's metabolism.

Enzymes are proteins; large three-dimensional amino acid polymers which adopt specific conformations (shapes) and recognise only certain substrates. The high specificity of substrate recognition is a consequence of a three-dimensional **'active site'** on each enzyme. An active site is a region in the three-dimensional structure that will accommodate only a particular shape and size of substrate. An analogy may be drawn with a 'lock and key' – the enzyme is the 'lock' and only one type of 'key', a specific substrate, will fit it!

It is relatively straightforward to list those properties of enzymes which make them good catalysts:

- Enzymes **bind** and hold 'still' their substrate (this is far more effective than relying on random collisions between molecules). This is often how chemical catalysts work, by binding and 'presenting' molecules or atoms on their surface.
- Enzymes make reactions feasible by providing an alternative reaction route with a lower activation energy.

Easy to say, but exactly how do enzymes **bind** their substrates, and how do they make reactions feasible?

Binding of substrate

The secret lies in the sequence of amino acids, linked together through peptide bonds, to form the polymeric enzyme (protein) molecule. The general structure of an amino acid is shown below.

$$H_2N-\underset{\displaystyle R}{\overset{\displaystyle H}{\underset{|}{\overset{|}{C}}}}-COOH$$

In a protein (polypeptide) chain, the amino (NH_2) and carboxyl (COOH) groups of each amino acid are involved in peptide bonding. The R group is specific for each amino acid.

Some R groups contain a carboxylate or an amino group which is charged (dissociated or protonated) at physiological pH; examples include:

- R group of glutamic acid, $HOOC(CH_2)_2$- dissociates to $^{-}OOC(CH_2)_2$-
- R group of lysine, $H_2N(CH_2)_4$- is protonated to $H_3{}^{+}N(CH_2)_4$-

Some R groups are polar, but not dissociated, because they contain a relatively highly electronegative atom, such as the sulphydryl group in cystine, $HS\text{-}CH_2$-

R groups may be apolar, such as the methyl groups in leucine, $(CH_3)_2\text{-}CH\text{-}CH_2$-

In addition to the various R groups, each peptide bond contains a carbonyl (-C=O) and an amine (-NH) group. These are all **functional** groups; many are dissociated and charged at physiological pH, or contain an electronegative atom and polar covalent bond. Arrangements of such groups about the three-dimensional 'surfaces' of the **active site** of an enzyme provide specific possibilities for charge–charge interactions or hydrophobic interactions with a substrate, which itself possesses interactive functional groups. The specific arrangement of amino acids at the active site leads to specific interactions with specific substrates. In other words, enzymes bind their substrates through intermolecular interactions. In Chapter 3 we learnt that such intermolecular interactions are relatively weak and reversible (as opposed to the intramolecular interactions which are strong covalent bonds). This is the basis of **binding**. Binding is a central concept in biology. Enzymes bind their substrates, hormones bind their receptors, antibodies bind antigens. Biological events, at the molecular level, are always preceded by binding, which results in a recognition and from which follows an action.

Providing an alternative reaction route

The second property of enzymes, namely that of providing an alternative reaction route with a lower activation energy, is also a property of the various R groups on the different amino acids. We can demonstrate this comprehensively by considering the mechanism of action of the enzyme **chymotrypsin**. Chymotrypsin is an enzyme which is present in our small intestine and which catalyses the hydrolysis of peptide bonds in proteins; in other words this enzyme helps to digest proteins into smaller peptides and amino acids which can be absorbed more readily into the body.

In the laboratory, the chemical hydrolysis of a peptide bond is a very slow process, unless a strong acid catalyst is added.

$$\mathrm{R{-}C({=}O){-}N(R){-}R} \; + \; \mathrm{H{-}O{-}H} \; \rightleftharpoons \; \mathrm{R{-}C({=}O){-}O{-}H} \; + \; \mathrm{R{-}N(R){-}H}$$

amide | water | carboxylic acid | amine

However, in the small intestine, where the pH is essentially neutral, the reaction catalysed by chymotrypsin occurs very quickly.

There are six stages in the mechanism of chymotrypsin catalysis which are explained in the following sequence.

Stage 1: A protein substrate approaches the active site of the enzyme. The active site of chymotrypsin allows that part of the enzyme to bind a portion of the protein which has a non-polar side chain, like those found in phenylalanine (consequently, those amino acid R groups lining this portion of the active site will be apolar). Once the protein is in place at the active site, an H^+ ion moves from the serine amino acid at position 195 (Ser-195) of the enzyme's amino acid sequence, to the histidine amino acid at position 57 (His-57). The oxygen atom in serine's hydroxyl group then forms a covalent bond to the carbon of one of the substrate's peptide bonds, in turn 'pushing' the two electrons from the pi bond on to the carbonyl oxygen. This is one transition state in the substrate transformation which is stabilised by the enzyme active site.

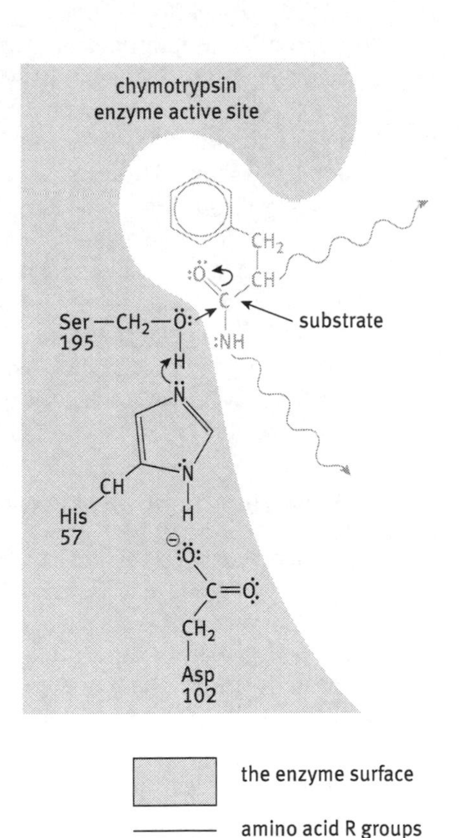

Stage 2: The positive charge formed on His-57 is stabilised by the negative charge on the aspartic acid at position 102 (Asp-102). When the double bond on the carbonyl group of the peptide bond is reformed, the bond between the carbon and the nitrogen in the peptide bond is broken. The nitrogen-containing group is stabilised by the formation of a bond to a hydrogen atom from His-57. This is a second transition state in the substrate transformation.

enzyme active site

Ser—CH₂—O
195

His
57

Asp
102

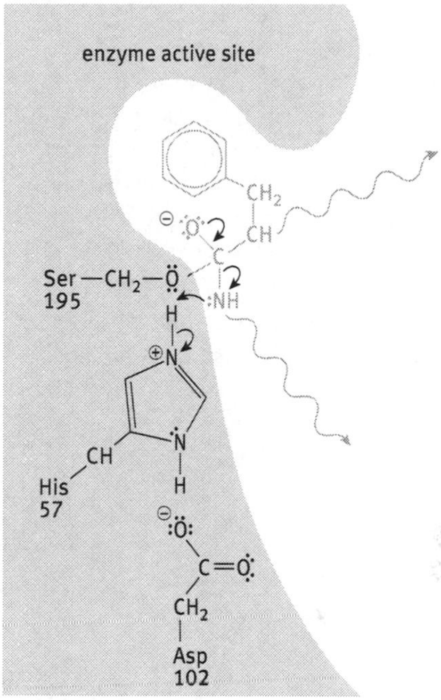

Stage 3: The portion of the polypeptide which contains the nitrogen atom from the broken peptide bond moves out of the active site. This represents a third transition state in the substrate transformation.

enzyme active site

Ser—CH₂—O
195

His
57

Asp
102

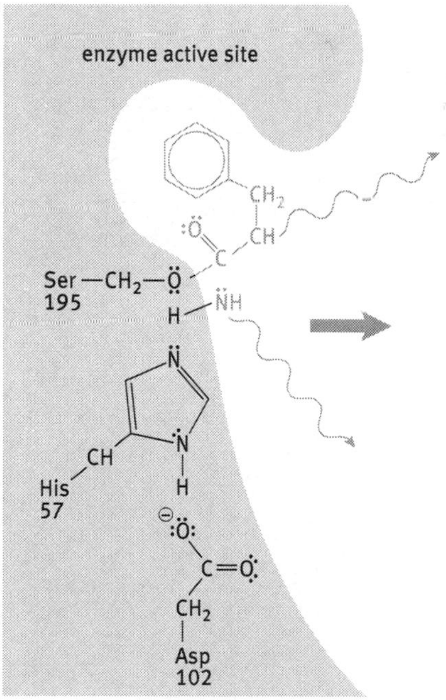

Stage 4: A water molecule moves into the active site. The oxygen atom in the water molecule loses an H^+ ion to a nitrogen atom on His-57. This allows water's oxygen atom to form a bond to the carbon atom on the remaining portion of the substrate. Like in stage 1, the pair of electrons from the pi bond moves to form a lone pair of electrons on the oxygen atom of the carbonyl group. Another transition state in the substrate is evident.

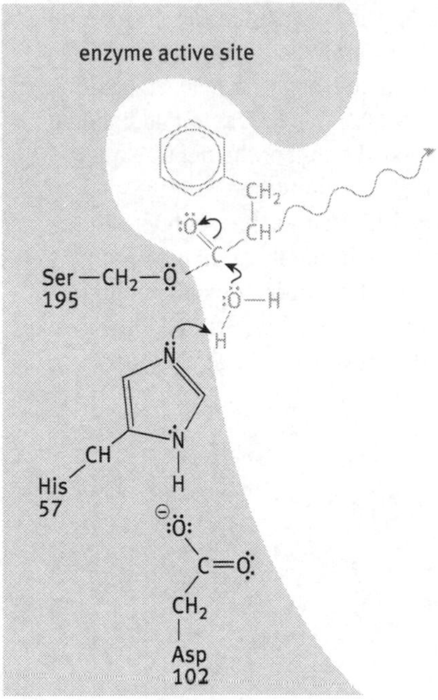

enzyme active site

Ser—CH_2—Ö
195
:Ö—H
H
N
CH
His
57
N
H
:Ö:
C=O
CH_2
Asp
102

Stage 5: When the carbonyl double bond reforms, the bond between carbon and the oxygen of Ser-195 is broken. The –OH group on Ser-195 is restored with a transfer of an H^+ ion from His-57. With this step, the Ser-195 and His-57 are both returned to their original forms.

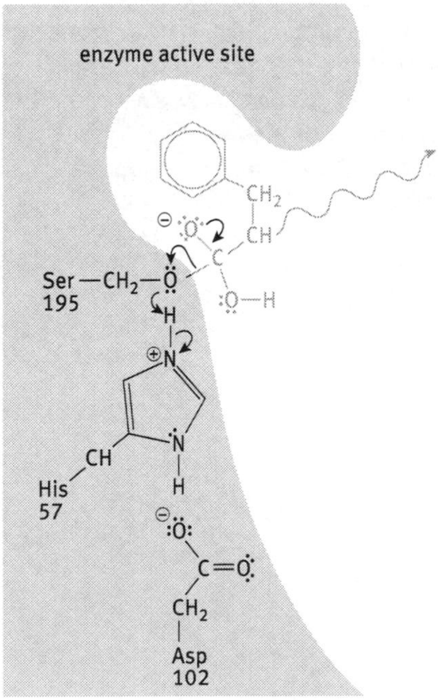

enzyme active site

Ser—CH_2—Ö
195
:O—H
H
⊕N
CH
His
57
N
H
:Ö:
C=O
CH_2
Asp
102

Stage 6: The remaining portion of the substrate moves out of the active site, leaving the active site in its original form, ready to repeat stages 1–5 with another protein molecule.

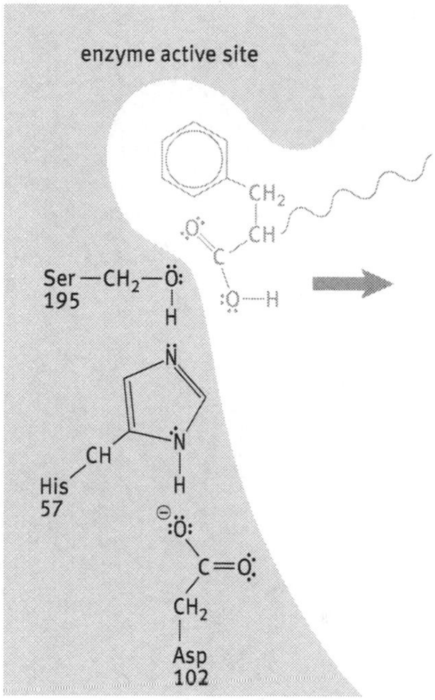

enzyme active site

Ser—CH_2—Ö:
195

His
57

Asp
102

Even this reaction scheme has been simplified! Nevertheless, it can be seen that a number of transition states must occur during this transformation, each being stabilised through the intervention of different functional groups in the enzyme active site.

So, functional groups on biomolecules such as proteins are responsible for specific binding events, mediate catalytic processes and, of course, make possible the intermolecular interactions within such molecules which define the molecules' active and functional three-dimensional shape.

Answers to "test yourself" questions

Answer 1.1
Hydrogen-1 (^1_1H) has just one proton and a mass number of 1. Hydrogen-2 (^2_1H), or deuterium, has an additional neutron and a mass number of 2 (1 proton + 1 neutron). Hydrogen-3 (^3_1H), or tritium, has two additional neutrons and a mass number of 3 (1 proton + 2 neutrons).

Answer 1.2
(a) Just the one, i.e. 1s.
(b) In energy level 2 we can find two types of atomic orbital, namely 2s and 2p; a total of four atomic orbitals may be present, namely $2s + 2p_x + 2p_y + 2p_z$.

Answer 1.3
A region in space about the atomic nucleus where there is a high probability of finding an electron.

Answer 1.4
(a) 10; two in 1s, two in 2s, and two in each of three 2p atomic orbitals.
(b) This would be an unreactive element since there are no unpaired electrons in the outer energy level which can participate in covalent or ionic bonding; the 2s and 2p atomic orbitals in energy level n = 2 are full (this element is in fact neon, one of the inert gases).

Answer 1.5
(a) 1s, 2s, ($2p_x$, $2p_y$, $2p_z$); all 2p atomic orbitals have equivalent energy.
(b) Atomic orbitals closest to the nucleus have lowest energy.

Answer 2.1
The atomic orbitals of two atoms merge to form one or more molecular orbitals in which electrons are shared in the formation of a covalent bond.

Answer 2.2
A sigma molecular orbital may be formed through the 'head-to-head' merging of s or p atomic orbitals, whereas a pi molecular orbital is formed by the 'side-to-side' merging of p atomic orbitals.

Answer 2.3
An 'asymmetric' covalent bond is one in which the electrons in the bond are drawn towards the more electronegative atom, creating a dipole (a separation of charge – hence the term 'polar'). The partial charges on the atoms allow for intermolecular charge–charge interactions with water.

Answer 2.4
Nothing really! A dative covalent (or coordinate) bond is the same as any other similar covalent bond, except for the fact that the two electrons constituting the bond are both provided by just one of the atoms in the bond, rather than the usual case of one electron being contributed by each of the two atoms forming the bond.

Answer 2.5
The 'octet rule' is explained in the 'Taking it further' section on the periodic table. For most of the lighter elements a complete outer, or valence, shell requires eight electrons, which leads to a more stable and unreactive state (Group 18 of the periodic table). These observations are said to obey the octet rule.

Answer 3.1
(a) A hydrogen bond (b) Common functional groups which participate in hydrogen bonding include hydroxyl (-OH) , carbonyl (-C=O), and amino (-NH_2).

Answer 3.2
The hydrogen bond is (i) a relatively strong form of intermolecular interaction, (ii) of a relatively 'fixed' distance, (iii) strongly directional.

Answer 3.3
Intermolecular interactions (i) are much weaker than covalent bonds, (ii) 'make and break' relatively quickly in water, (iii) tend to be rather dependent on pH.

Answer 3.4
Van der Waals short range intermolecular interactions.

Answer 3.5
(c) and (e); in (c) the methyl group (CH_3) shows no polarity since C and H have similar electronegativities, in (e) the carbon ring structure is particularly hydrophobic. The ring structure in (f) is made polar by the attachment of a hydroxyl group.

Answer 4.1
Remember, no. of moles = mass in grams/molar mass. Therefore, we have $0.01/6000 = 1.66 \times 10^{-6}$, or 1.66 micromoles of insulin.

Answer 4.2
The M_r of ethanoic acid is $(12 \times 2) + (4 \times 1) + (16 \times 2) = 60$. In other words, 60 grams of ethanoic acid is 1 mole. A 1 M solution of ethanoic acid would contain 1 mole (60 grams) in 1 litre. A 0.1 M solution of ethanoic acid would contain 1/10 moles (6 grams). It therefore follows that to make 10 litres of a 0.1 M solution of ethanoic acid, you would need $10 \times 6 = 60$ grams.

Answer 4.3
10 grams of glucose is equivalent to 10/180 = 0.055 moles. There are therefore 0.055 moles of glucose present in the 50 ml of solution. The molarity is defined as the number of moles in a litre. If there are 0.055 moles in 50 ml, then by extrapolation there would be $0.055 \times 1000/50$ (assuming no dilution) in a litre = 1.1 M (remember, the molarity does not change with volume [unless diluted], but the number of moles will).

Answer 4.4
If the stock solution of glycine is 0.02 M, then by definition this would contain 0.02 moles of glycine in 1 litre. Therefore, 1 ml of this solution must contain $0.02/1000 = 2 \times 10^{-5}$ moles. Of course, 1 ml of this solution is still 0.02 M! The total volume of the enzyme assay is 3 ml, 1 ml of which is from the added glycine solution. In this total volume there would still be 2×10^{-5} moles of glycine, but the concentration has been diluted. The molarity is now (1 in 3 dilution) $0.02/3 = 0.0066$ M (6.6×10^{-3} M).

Answer 4.5
1.2 grams of glycine is equivalent to 1.2/79 = 0.015 moles. (a) In 1 ml of this 100 ml there must be 0.015/100 = 0.00015 moles. (b) One litre of this solution would contain 1000×0.00015 moles = 0.15 mol l^{-1}, i.e. a 0.15 M solution. (c) 1 ml of this solution must contain $0.00015 \times 6.022 \times 10^{23}$ (since 1 mol is equivalent to 6.022×10^{23} molecules = Avogadro's number) = 9.033×10^{19} molecules.

Answer 5.1
In an sp^3 hybridised carbon atom, there are four hybridised atomic orbitals derived from the 2s and three 2p atomic orbitals. In an sp^2 hybridised carbon, there are three hybridised atomic orbitals leaving one 2p atomic orbital unchanged.

Answer 5.2
In methane the carbon is sp^3 hybridised; its four hybridised atomic orbitals each produce a covalent bond with hydrogen. The four C-H bonds arrange themselves as far apart from one another as possible; consequently the three-dimensional shape of methane is that of a tetrahedron.

Answer 5.3
In (a), (b) and (d) the carbon must be forming four sigma covalent bonds, i.e. the carbon is sp^3 hybridised.

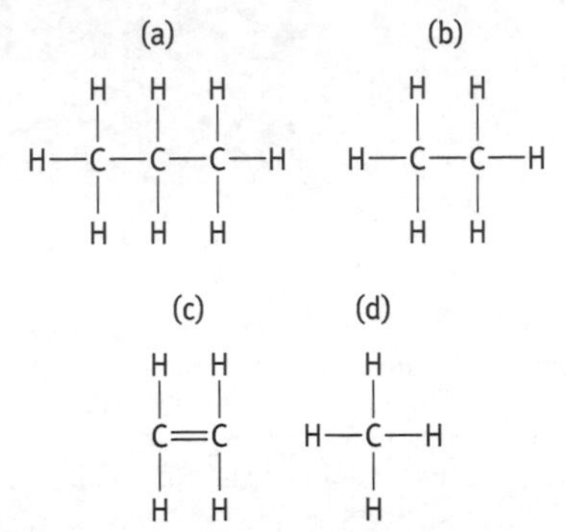

In (c) however, the only way that carbon could satisfy its valency and form four covalent bonds is if it were sp^2 hybridised and therefore able to form a pi bond through the unhybridised p orbital, so creating a carbon–carbon double bond.

Answer 5.4
B and C. In B, four of the six carbons are sp^2 hybridised (those either side of the two double bonds), while in C every carbon is sp^2 hybridised and the ring is aromatic. In A, each carbon is sp^3 hybridised, forming four covalent bonds; one to each of two adjacent carbons, and two to each of two hydrogens (not shown).

Answer 6.1
(i); the two molecules are structural isomers.

Answer 6.2
Both pairs of molecules are stereoisomers, but enantiomers are mirror images of each other; diastereomers are not mirror images.

Answer 6.3
A chiral centre (most often an asymmetric carbon atom) has four different groups attached to it.

Answer 6.4
The number of possible stereoisomers is given by 2^n, where n is the number of chiral centres. The answer is therefore $2^6 = 64$.

Answer 6.5
By observing their effect on the rotation of plane polarised light. D-glucose will rotate plane polarised light to the right (it is dextrorotatory), whereas L-glucose will rotate it to the left (it is levorotatory).

Answer 7.1
Water is a dipolar molecule. The electronegative oxygen atom draws electrons to itself, causing the hydrogen atoms to have a partial positive charge. In the diagram as shown the water molecules would repel each other.

Answer 7.2
(a) An acid is a substance which produces H^+ ions by dissociation
(b) A strong acid will essentially completely dissociate in solution (its dissociation constant K_a will be very large), whereas a weak acid will only partially dissociate and its K_a will be smaller
(c) The pH scale is a log scale, from 0 to 14
(d) Remember, each pH unit represents a 10-fold increase (or decrease) in $[H^+]$; so an increase in pH from 9 to 4 is five pH units, which equals a $10 \times 10 \times 10 \times 10 \times 10$, i.e. a 100 000 fold increase in $[H^+]$.

Answer 7.3
(a) Assuming the strong acid is completely dissociated, then $[H^+] = 0.05$ M, so using the equation:
$pH = -\log[H^+]$
$= -\log[0.05]$
$= -(-1.30)$
$= 1.30$
(b) For a solution of pH 6.2, the $[H^+]$ is given by:
$6.2 = -\log[H^+]$
the negative antilog of 6.2 is 6.3×10^{-7} [*on the calculator enter 'shift' 10^x-6.2*]
Therefore, $[H^+] = 6.3 \times 10^{-7}$ M

Answer 7.4
$pK_a = -\log_{10}K_a$
therefore,
$pK_a = -\log_{10}(1.8 \times 10^{-5})$
$= -(-4.74)$
$= 4.74$

Answer 7.5
Use the Henderson–Hasselbalch equation.

$$pH = pK_a + \log\frac{[base]}{[acid]}$$
$$= 4.75 + \log\frac{0.05}{0.10}$$
$= 4.75 + (-0.30)$ [*enter log (0.05 divided by 0.10)*]
$= 4.45$

Answer 7.6
Using the Henderson–Hasselbalch equation:

$$pH = pK_a + \log\frac{[base]}{[acid]}$$
$$= 8.08 + \log\frac{0.186}{0.14}$$
$= 8.08 + (0.123)$
$= 8.20$

Answer 7.7
That molecule with a pK_a of 4.2 would give the most acid solution. Using the equation $pK_a = -\log_{10}[K_a]$, i.e. taking the negative antilog of each pK_a value, the $[H^+]$ are respectively 6.31×10^{-5}, 1.58×10^{-7} and 6.31×10^{-9} M.

Answer 8.1
(c) is correct; there are no heat gradients in cells to do work.

Answer 8.2
(b) is correct; this is the only catabolic process, all the other processes are synthetic (anabolic) and will require a net input of energy to make them feasible.

Answer 8.3
(c) is correct; lowering the activation energy.

Answer 8.4
Exergonic describes reactions/processes which result in the release of free energy, i.e. a negative ΔG. Endergonic processes require an input of free energy (ΔG is positive), usually through being coupled to an exergonic process.

Answer 8.5
Firstly, to calculate the entropy change, use the equation:
$\Delta G = \Delta H - T\Delta S$
$-3089.0 \text{ kJ mol}^{-1} = -2807.8 \text{ kJ mol}^{-1} - (310 \times \Delta S)$ [*T=273 + 37 =310 K*]
To simplify, if we add 2807.8 kJ mol^{-1} to both sides, we get:
$-281.2 \text{ kJ mol}^{-1} = -310 \text{ K} \times \Delta S$

$$\Delta S = \frac{-281.2 \text{ kJ mol}^{-1}}{-310 \text{ K}}$$ [*dividing two minuses gives a plus!*]

$= 0.907 \text{ kJ mol}^{-1} \text{ K}^{-1}$
The thermodynamic data provided for this reaction would suggest it is spontaneous, i.e. ΔG is negative, as is ΔH (reaction is exothermic). Furthermore, the calculated ΔS is positive. We would assume the reaction to be feasible and both 'enthalpy-driven' and 'entropy-driven'.

Answer 9.1
The equilibrium constant, K_{eq}, is given by the expression:

$$K_{eq} = \frac{[\alpha\text{-ketoglutarate}][CO_2][NADH]}{[isocitrate][NAD^+]}$$

Answer 9.2
An equilibrium constant of 3.18×10^{520} (which is very high!) would indicate that the reaction goes almost to completion with almost no reactants left.

Answer 9.3

Using the equation:

$\Delta G = -RT \ln K_{eq}$, we have

$-3.7 \times 10^3 \text{ Jmol}^{-1} = -8.314 \text{ JK}^{-1}\text{mol}^{-1} \times 310 \text{ K} \times \ln K_{eq}$

(multiplying ΔG *by* 10^3 *because R is in Joules and* ΔG *is in kilojoules; and using the absolute temperature T as 273 + 37 = 310 K)*

therefore,

$$\ln K_{eq} = \frac{-3.7 \times 10^3}{-8.314 \times 310}$$

$= 1.44$

and so $K_{eq} = 4.20$ M *(use 'shift'* e^x *on the calculator to give e1.44 = 4.20)*

ΔG is relatively small for this reaction, and the value of K_{eq} is likewise low, suggesting that there are significant amounts of reactant still present.

Answer 9.4

Using the equation:

$\Delta G^{o'} = -RT \ln K_{eq}$

and since *R* (the gas constant) is given in joules, substituting values gives:

$13000 \text{ Jmol}^{-1} = 8.314 \text{ JK}^{-1}\text{mol}^{-1} \times 310 \text{ K} \times \ln K_{eq}$

$$\ln K_{eq} = \frac{-13\,000}{-8.314 \times 310}$$

$\ln K_{eq} = -5.04$

$K_{eq} = 0.00645 = 6.45 \times 10^{-3}$M

Answer 10.1

The two half-reactions for (a) are: $Zn \rightarrow Zn^{2+} + 2e^-$ and $Cu^{2+} + 2e^- \rightarrow Cu$, and for (b) are: $Fe^{2+} \rightarrow Fe^{3+} + e^-$ and $Cu^{2+} + e^- \rightarrow Cu^+$

Answer 10.2

(a) The two half-reactions are written as:

acetaldehyde + $2H^+ + 2e^- \rightarrow$ ethanol

$NADH + H^+ \rightarrow NAD^+ + 2e^- + 2H^+$

(b) Using the tabulated redox potential values given in the text, for acetaldehyde/ethanol, $\Delta E^{o'} = -0.2$ V and for NAD^+ / NADH. $\Delta E^{o'} = -0.32$ V. However, in the reaction as shown, NADH is being oxidised; therefore we reverse the sign of the redox potential (which by convention is given as a reduction reaction), and so the $\Delta E^{o'}$ is +0.32 V.

Remember that electrons will tend to flow from a 'more negative' to a 'more positive' redox potential. Here we have electrons flowing from the oxidation of NADH ($\Delta E^{o'} = 0.32$ V) to the reduction of acetaldehyde ($\Delta E^{o'} = -0.2$ V). The reaction therefore is favoured and will tend to proceed from left to right.

Answer 10.3

Using tabulated redox values, $\Delta E^{o'}$ for reduction of ubiquinone (ubiquinone $_{(oxidised)} \rightarrow$ ubiquinone $_{(reduced)}$) is 0.10 V, and $\Delta E^{o'}$ for reduction of cytochrome *c* (cytochrome *c* $_{(oxidised)} \rightarrow$ cytochrome *c* $_{(reduced)}$) is 0.254 V. The reaction as written shows a reduction of ubiqinone ($\Delta E^{o'} = 0.10$ V), electrons being supplied by the oxidation of cytochrome *c* (so $\Delta E^{o'} = -0.254$ V, reversing the 'sign'). Since electrons tend to flow from a 'more negative' to a 'more positive' redox potential, this reaction is likely to proceed from right to left; therefore the reaction as shown is unfavourable.

Answer 10.4

Using tabulated redox potential values; succinate is being oxidised (donating two electrons to FAD), therefore the $\Delta E^{o'}$ is −0.03 V. For FAD, which is being reduced, $\Delta E^{o'} = -0.18$ V. The change in redox potential ($\Delta E^{o'}$) is therefore: $-0.03 + (-0.18) = -0.21$ V. Inserting this value into the equation gives:

$\Delta G^{o'} = -nF\Delta E^{o'}$

$\Delta G^{o'} = -2 \times 96485 \times -0.21$

$\Delta G^{o'} = +40524 \text{ J mol}^{-1}$ *(remember, two minuses make a plus)*

The $\Delta G^{o'}$ value for this reaction as written is positive, suggesting the reaction is unfavourable and is more likely to proceed from right to left.

Answer 10.5

In this reaction as written, NAD^+ is being reduced to NADH ($\Delta E^{o'} = -0.32$ V), and malate is being oxidised to oxaloacetate ($\Delta E^{o'}$ is therefore +0.17 V). The $\Delta E^{o'}$ for this reaction is therefore $-0.32 + 0.17 = -0.15$ V. Inserting this into the equation gives:

$\Delta G^{o'} = -nF\Delta E^{o'}$

$\Delta G^{o'} = -2 \times 96485 \times -0.15$

$\Delta G^{o'} = 28945 \text{ J mol}^{-1}$ *(remember, two minuses make a plus)*

A positive $\Delta G^{o'}$ suggests this reaction as written is unfavourable and is more likely to proceed from right to left.

Answer 11.1

A nucleophilic site on a molecule is an area which is 'electron rich' and one which will attract a nucleophile, such as a proton. Nucleophilic sites may result from electronegative atoms forming polar covalent bonds, pi bonds, lone electron pairs on an atom, or dissociated groups. An electrophilic centre is one which carries a positive charge and therefore attracts an electrophile.

Answer 11.2
Pi bonds, which are formed by the side-to-side merger of p atomic orbitals, effectively produce an 'electron-rich cloud' above and below the plane of the sigma covalent bond. The two electrons in the pi bond readily 'add' to an electrophile in an addition-type reaction.

Answer 11.3
The hydrocarbon molecule (ethene) contains a carbon–carbon double bond. In the molecule A-B, B is a more electronegative atom, indicated by the negative charge on B, the positive charge on A, and the proximity of the two electrons (··) in the bond to B. A is therefore an electrophile and is attracted to the pi bond. As A-B approaches the pi bond, electrons in the A-B bond are pushed further onto atom B (shown by the small arrow in the diagram). Eventually, the A-B bond becomes so polarised that it breaks; A forms a new bond to carbon using the two electrons in the pi bond (a form of coordinate bonding). This leaves a positive charge on the other carbon atom (now an electrophilic site) which attracts the ion B^- with its lone pair of electrons. B 'adds' to C, providing the two electrons needed to form a new covalent bond.

Answer 11.4
Enzymes are biological catalysts which make reactions feasible by providing alternative reaction routes with lower activation energies. Intermolecular interactions at the active site of the enzyme with its substrate are able to stabilise different transition states which are the basis of the alternative reaction route (with a lower activation energy).

Answer 11.5
When covalent bonds are broken, they normally do so in a heterolytic fashion, i.e. the two electrons in the bond go to one or other of the two atoms involved; one atom will effectively gain an electron (becoming negatively charged) while the other atom effectively loses an electron, becoming positively charged. Again, two ions are formed, although they are likely to exist only transiently. In both cases the normal event is for the two electrons comprising the covalent or ionic bond to go to one or other of the two atoms when the bond is broken. However, a free radical is formed when a covalent bond is broken homolytically, i.e. the bond is split equally such that the two electrons in the bond are shared equally between the two atoms. This produces a free radical, containing an atom with a single unpaired electron. This is an unstable state and that atom will be very reactive, looking for an additional electron to pair with. It will often gain this electron by 'stealing' from another atom, thereby generating another radical, then another, in a chain reaction. This can cause serious damage to biological molecules.

Appendix 1. Some common chemical formulae

Formula	Name	Type of compound
HCl	Hydrochloric acid	Strong acid in aqueous solution
H_2SO_4	Sulphuric acid	Strong acid in aqueous solution
HNO_3	Nitric acid	Strong acid in aqueous solution
H_3PO_4	Phosphoric acid	Weak acid in aqueous solution
NH_3	Ammonia	Weak base in aqueous solution
CO_2	Carbon dioxide	Gas at STP (standard temperature and pressure) Weak acid in aqueous solution
CO	Carbon monoxide	Gas at STP
CH_4	Methane	Alkane
CH_3OH	Methanol	Alcohol
HCHO	Methanal	Aldehyde
HCOOH	Methanoic (formic) acid	Carboxylic acid (weak acid)
CH_3CH_3	Ethane	Alkane
CH_2CH_2	Ethene	Alkene
C_6H_6	Benzene	Aromatic hydrocarbon
CH_3CH_2OH	Ethanol	Alcohol
CH_3CHO	Ethanal	Aldehyde
CH_3COOH	Ethanoic (acetic) acid	Carboxylic acid (weak acid)

Appendix 2. Common anions and cations

Cations		Anions	
Formula	**Name**	**Formula**	**Name**
Na^+	Sodium	F^-	Fluoride
K^+	Potassium	Cl^-	Chloride
Ca^{2+}	Calcium	Br^-	Bromide
Mg^{2+}	Magnesium	SO_4^{2-}	Sulphate
Fe^{2+}	Iron(II) - ferrous	CO_3^{2-}	Carbonate
Fe^{3+}	Iron(III) - ferric	NO_3^-	Nitrate
NH_4^+	Ammonium	OH^-	Hydroxide
		HCO_3^-	Hydrogen carbonate
		PO_4^{3-}	Phosphate
		HPO_4^{2-}	Hydrogen phosphate
		$H_2PO_4^-$	Dihydrogen phosphate

Appendix 3. Common functional groups

Group	Name
$>C=C<$	Alkene (pi double bond)
$-O-H$	Hydroxyl
$>C=O$	Carbonyl (ketone)
$-C(=O)H$	Aldehyde
$-C(=O)OH$	Carboxyl
$-NH_2$	Amine
$-C(=O)NH_2$	Amide
$-C(=O)N(R)-$	Peptide linkage
R	General representation of an alkyl group (C_nH_{2n-1})

Appendix 4. Notations, formulae and constants

Atoms

${}^{A}_{Z}X$ where A = the mass number and Z = the atomic number

Molecules

A line connecting atoms in a structural formula denotes a covalent (sigma bond), a double line denotes a double bond, i.e. a sigma bond plus a pi bond. A solid wedge denotes a bond coming out of the page, a hatched wedge denotes a bond going in to the page.

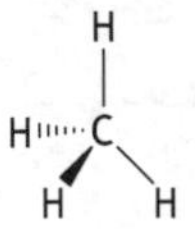

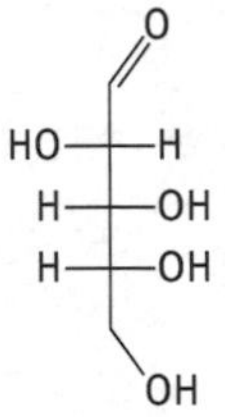

can be written as

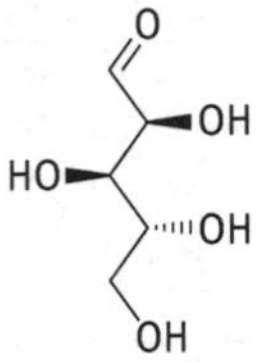

Carbon atoms are not normally shown in chain or ring structures, nor indeed the hydrogen atoms attached to them.

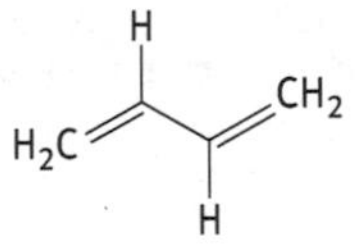

can be written as

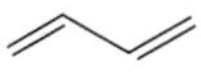

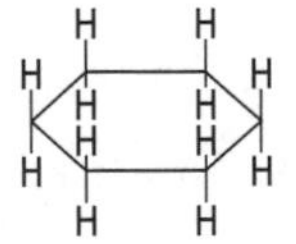

can be drawn as

Atoms other than carbon, that form part of a ring structure, are always shown.

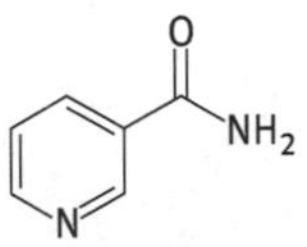

Amounts and concentrations

Avagadro's number (constant) = 6.022×10^{23}, the number of atoms, molecules or particles in 1 mole of a substance. A mole is an amount; the amount that contains a number of molecules equal to Avagadro's number.

One **mole** (mol) of a compound is also the amount of the substance equal to its molecular mass expressed in grams. For example, the molecular formula of glucose is $C_6H_{12}O_6$. Its molecular mass is given by the sum of the atomic masses of its atoms, which is approximately (6×12) + (12×1) + 6×16) = 180; therefore we can say that 180 g of glucose is equivalent to 1 mole.

Molarity (M) is a concentration; 1 mole of a solute dissolved in 1 litre of solvent makes a 1 M solution. In other words, 180 g of glucose dissolved in 1 l of water is a 1 M glucose solution. If we take 1 ml of that solution we have not changed the concentration (the ratio of glucose to water is the same), it is still 1 M, but that 1 ml only contains 1 mmol, one thousandth of a mole.

Acids and bases

Acidity is measured according to the hydrogen ion concentration, denoted by $[H^+]$ or $[H_3O^+]$. Hydrogen ion concentrations ($[H^+]$) are generally very small and so the pH scale is used to denote hydrogen ion concentration on a scale of 0–14.

$$pH = -\log_{10}[H^+]$$

Water dissociates weakly (the auto-ionisation of water) to give a neutral solution.

K_w is the dissociation constant of water:

$$K_w = [H^+][OH^-] = [1 \times 10^{-7}][1 \times 10^{-7}] = 1 \times 10^{-14}\ M^2\ (\text{or mol}^2\ \text{dm}^{-6})$$

Acids dissociate in solution; a weak acid dissociates partly and a strong acid almost completely: K_a = acid dissociation constant for the reaction.

In $AH + H_2O \rightleftharpoons A^- + H_3O^+$ (the acid AH dissociates to form its conjugate base, A^-)

$$K_a = \frac{[A^-][H_3O^+]}{[AH]}$$

Values of acid dissociation constants are usually very small and so, as in the case of the hydrogen ion concentration, we take the log and express the result as a pK_a.

$$pK_a = -\log_{10}K_a$$

To calculate the pH of a solution of a weak acid in equilibrium with its conjugate base we can use the Henderson–Hasselbach equation:

$$pH = pK_a + \log_{10}\frac{[\text{base}]}{[\text{acid}]}$$

where [base] = concentration of conjugate base and [acid] = concentration of conjugate acid.

This equation also allows us to calculate the pH of buffer solutions.

Thermodynamics

Standard Conditions: The symbol Δ is used to denote a change in a quantity such as the enthalpy or entropy of a system. This is because we cannot measure absolute values but only a change in such quantities. In order to ensure we are comparing like with like we need to define the conditions under which we are measuring these changes to ensure they are similar throughout a reaction or other change. The reference conditions which chemists use to compare these changes are called the conditions of standard-state and are represented by the superscript ° following the quantity. For example ΔH° means the enthalpy change when reactants in their standard states are converted to products in their standard states. The standard state of a substance is the pure form of the substance at a pressure of 1 atm pressure. For solutions this refers to a 1 M concentration of the substance. Standard enthalpy changes are usually reported at a temperature of 25°C or 298 K. There is one further condition which is important for biological changes and this is the pH. Biologists use the symbol $\Delta G^{\circ\prime}$ to denote a free energy change under biological conditions which defines the pH as being 7.0.

ΔH = change in enthalpy (kJ mol^{-1}).

ΔS = change in entropy (kJ mol^{-1}K^{-1}).

ΔG = change in Gibbs free energy (kJ mol^{-1}).

Gibbs equation is $\Delta G = \Delta H - T\Delta S$, where T is the absolute temperature in Kelvin (K).

ΔG° = the standard free energy change at 1 atmosphere and 298 K.

$\Delta G^{\circ\prime}$ = the standard free energy change which refers to the standard state in an ideal solution (commonly used for biological systems and measured at 25°C, pH = 7.0, 1 atmosphere pressure and 1 M concentration).

A positive ΔG indicates an endergonic process.

A negative ΔG indicates an exergonic process.

A positive ΔH indicates an endothermic process.

A negative ΔH indicates an exothermic process.

Kinetics

The rate of a reaction is given by the change in concentration of the reactant [R] with time:

$$\text{rate} = \frac{-\Delta[\text{R}]}{\Delta t}$$

Or the rate of increase in the concentration of the product [P]:

$$\text{rate} = \frac{\Delta[\text{P}]}{\Delta t}$$

For the reaction A + B → C + D:

$$\text{rate} = k[\text{A}]^x[\text{B}]^y$$

where k is the rate constant (for a stated temperature), and x and y are orders of reaction.

If x and y are zero, i.e. the reaction rate does not depend on the concentration of any of the reactants (common for enzyme-catalysed reactions), then the rate of the reaction is equal to the rate constant:

rate = k.

This arrow symbol $\rightleftharpoons$ denotes a reversible reaction.

At equilibrium there is no net change in the concentration of products or reactants; the forward and reverse rates are equal.

For the reaction:

$$\text{pyruvate} + \text{NADH} + \text{H}^+ \rightleftharpoons \text{lactate} + \text{NAD}^+$$

the equilibrium constant, K_{eq}, is given by the expression:

$$K_{eq} = \frac{[\text{lactate}][\text{NAD}^+]}{[\text{pyruvate}][\text{NADH}][\text{H}^+]}$$

Free energy and equilibrium

Free energy and equilibrium constants are related by the expression:

$$\Delta G^{o\prime} = -RT \ln K_{eq}$$

where R is the gas constant and T the absolute temperature.

Free energy and redox potential

Free energy and redox potentials are related by the Nernst equation:

$$\Delta G^{o\prime} = -nF\Delta E^{o\prime}$$

where F is a constant, the Faraday, n is equal to the number of electrons transferred in the reaction and $E^{o\prime}$ is the redox potential (V).

Reactivity

In describing reaction mechanisms, a short 'full head' curly arrow is used to indicate the movement of a pair of electrons. A 'half-head' curly arrow describes the movement of a single electron (with the formation of a radical).

for **pairs** of electrons
(more common)

for a **single** electrons
(*i.e.* radical reactions)

Two 'dots' against an atom denotes a lone pair of electrons. For example, oxygen (in the carbonyl group) has two lone pairs of electrons:

$>C=\ddot{\underset{..}{O}}$

One 'dot' against an atom denotes a single unpaired electron (a radical).

Cl· is the chlorine radical.

Appendix 5. Glossary

Acid (Brønsted-Lowry definition): a substance which can donate a proton or hydrogen ion.

Acid dissociation constant (K_a): the equilibrium constant for the dissociation of a proton from an acid.

Activation energy: the amount of energy that reactants must absorb before a chemical reaction will start (to promote the reactants from the ground state to the transition state).

Active site: the area of an enzyme where the substrate binds by means of weak chemical bonds. It is often located in a cleft or pocket in the protein's tertiary structure.

Active transport: the process by which dissolved molecules move across a cell membrane from a lower to a higher concentration. In active transport, particles move against the concentration gradient and either maintain or increase the concentration gradient.

Addition reaction: the addition of a small molecule (e.g. H_2) to a double or triple carbon bond.

Alcohol: an organic molecule containing a –OH group attached to a carbon atom that is not part of a carbonyl group or an aromatic ring.

Aldehyde: an organic molecule containing the –CHO group.

Amine: an organic molecule derived from ammonia by replacing various numbers of H atoms with organic groups. Amines contain the group $-NH_2$, -NH or –N.

Amino acid: an organic molecule possessing both carboxyl and amino groups. Amino acids serve as the monomers of proteins.

Amino group: a functional group that consists of a nitrogen atom bonded to two hydrogen atoms. It can act as a base in solution, accepting a hydrogen ion and acquiring a charge of +1.

Amphipathic (also amphiphilic): a molecule that has both a hydrophilic region and a hydrophobic region.

Anabolic pathway: a metabolic pathway that synthesises a complex molecule from simpler compounds.

Angstrom (Å): a unit of length equal to 1×10^{-10} m.

Anion: a negatively charged ion.

Anomeric carbon: the most oxidised carbon atom of a cyclised monosaccharide. The anomeric carbon has the chemical reactivity of a carbonyl group.

Anomers: isomers of a sugar molecule that have different configurations only at the anomeric carbon atom.

Apolar: opposite of polar. See *Polar*.

Aromatic hydrocarbon: an aromatic hydrocarbon is one which incorporates one or more planar sets of six carbon atoms that are connected by delocalised electrons numbering the same as if they consisted of alternating single and double covalent bonds. The simplest aromatic hydrocarbon is benzene; this configuration of six carbon atoms is known as a benzene ring.

Asymmetric carbon: a carbon atom which is attached to four different atoms or groups.

Atom: the smallest unit of matter that retains the properties of an element.

Atomic mass unit (amu): the unit of atomic mass equal to one-twelfth the mass of the ^{12}C isotope of carbon.

Atomic nucleus: an atom's central core that contains protons and neutrons.

Atomic orbital: a region in space around an atom in which there is a high probability of finding an electron.

ATP: adenosine triphosphate. An adenosine-containing nucleoside triphosphate that releases free energy when its phosphate bonds are hydrolysed. The 'energy currency' of the cell. The free energy released is used to drive endergonic reactions in the cell.

Auto-ionisation of water: a reaction in which a proton is transferred from one molecule of water to another to form an H_3O^+ ion and an OH^- ion.

$$2H_2O \rightleftharpoons H_3O^+ + OH^-$$

Autotroph: autotrophic organisms use energy from the sun or from the oxidation of inorganic compounds to make organic molecules.

Avagadro's number: formally defined as the number of carbon-12 atoms in 0.012 kg of carbon-12. Avogadro's number is 6.022×10^{23}.

Base (Brønsted-Lowry definition): a substance which can extract a proton, e.g. OH^-, NH_3, RNH_2.

Beta-pleated sheet: one form of the secondary structure of proteins.

Bond energy: an average value for the energy required to break a covalent bond in such a way that each participating atom retains an unpaired electron.

Boyle's law: states that at a constant temperature the volume of a fixed amount of gas varies inversely with its pressure.

Buffer: a substance that consists of conjugate acid and base forms in a solution that minimises changes in pH when extraneous acid or base is added to the solution.

Calorie: the amount of heat energy required to raise the temperature of 1 g of water by 1°C. The Calorie, with a capital C, usually used to indicate the energy content of food, is a kilocalorie.

Carbohydrates: a sugar (monosaccharide), or one of its dimers (disaccharide) or polymers (polysaccharide), in which the ratio C: H: O is usually 1: 2: 1.

Carbonyl group: the functional group >C=O found in aldehydes and ketones.

Carboxylate ion: the conjugate base of a carboxylic acid with general formula $RCOO^-$.

Carboxyl group: the functional group –COOH found in carboxylic acids.

Carboxylic acid: an organic molecule containing the –COOH group. They are weak acids.

Carrier protein: a protein which can transport molecules across a cell membrane usually by adopting different conformations to accept, transport and release the molecule.

Catabolic pathway: a metabolic pathway that releases energy by breaking down complex molecules to simpler compounds.

Catalyst: a chemical agent that changes the rate of a reaction without itself being consumed in the reaction.

Cation: an ion with a positive charge, produced by the loss of one or more electrons.

Channel protein: a protein which can transport a molecule across a cell membrane by adopting open and closed conformations to provide a continuous pathway.

Charles's law: states that at constant pressure the volume of a fixed amount of gas is directly proportional to its absolute temperature.

Chemical bond: an attraction between two atoms resulting from a sharing of outer energy level electrons, or the presence of opposite charges on ions.

Chemical energy: energy stored in the chemical bonds of molecules; a form of potential energy.

Chemical equilibrium: a stage in a chemical reaction in which the rate at which the reactants are converted to products is equal to the rate at which products are being converted back to reactants. A chemical equilibrium is a dynamic equilibrium.

Chemical reaction: a process that results in chemical changes to matter, and involves the making and/or breaking of chemical bonds.

Chemo-autotroph: an organism that needs only carbon dioxide as a carbon source but that obtains energy by oxidising inorganic compounds.

Chiral (or asymmetric) molecule: a molecule which cannot be superimposed upon its mirror image.

Cholesterol: a steroid that forms an essential component of animal cell membranes and which acts as a precursor for a number of other biologically important steroids.

Concentration gradient: the unequal difference in concentration of ions across a cell membrane. Any difference in the concentration of a solute between two regions within a solution.

Condensation reaction: a reaction in which two molecules become covalently bonded to each other through the loss of a small molecule, usually water (a dehydration reaction).

Coordinate bond: a chemical bond formed by the sharing of a pair of bonding electrons between two atoms, when the electron pair originates from one of the atoms. See also *Dative covalent bond*.

Covalent bond: a chemical bond formed by the sharing of a pair of electrons.

Conjugate acid/base: the product resulting from the gain of a proton by a base; or, the product resulting from the loss of a proton by an acid.

Conjugated system: a chain of atoms (usually carbon) connected by alternating single and double bonds.

Cytochrome: an iron-containing protein component of electron transport chains in mitochondria and chloroplasts.

Dalton: a unit of atomic mass equivalent to one-twelfth of the mass of the atom ^{12}C.

Dative covalent bond: a chemical bond formed by the sharing of a pair of bonding electrons between two atoms, when the electron pair originates from one of the atoms. See also *Coordinate bond*.

Dehydration reaction: a chemical reaction in which two molecules covalently bond to each other with the removal of a water molecule.

Delocalisation of electrons: the spread of electrons over several atoms within a molecule.

Deoxyribonucleic acid (DNA): a double-stranded helical nucleic acid molecule that is capable of replicating and determining the inherited structure of a cell's proteins.

Deoxyribose: the sugar component of DNA, having one less hydroxyl group than ribose, the sugar component of RNA.

Diffusion: a reduction in concentration gradient due to the random motion of particles.

Dipole: two equal but opposite charges, separated in space, resulting from the unequal distribution of charge within a molecule or chemical bond. See also *Polar.*

Dissociation:
i. breaking of a bond
ii. separation into ions by an ionic compound on dissolving in water.

Dissociation constant (of an acid), K_a: the equilibrium constant which indicates the degree to which an acid is dissociated into ions in water.

Electron: a negatively charged sub-atomic particle found outside the nucleus of an atom.

Electron configuration: the way in which the electrons of an atom are arranged in orbitals in an atom or molecule, e.g. C $1s^2 2s^2 2p^2$

Electron energy level: a permitted value for the energy of an electron in an atom or molecule.

Electron shell: 'shell' is used interchangeably with energy level. See *Electron energy level.*

Electron transport chain: a sequence of electron carrier molecules that possess redox components and collectively shuttle electrons along their length.

Electrophile: species that are attracted to negative centres. Electrophiles are typically cations such as the hydroxonium ion, H_3O^+.

Electronegativity: the attraction of an atom for the electrons in a covalent bond.

Electrostatic interactions: a general term for the electronic interaction between particles. Electrostatic interactions include charge–charge interactions, hydrogen bonds, and van der Waals forces.

Element: a substance which cannot be separated into simpler substances by chemical techniques. All atoms of the same element have the same atomic number and electron configuration.

Elimination reaction: a chemical reaction in which a molecule forms two different molecules.

Enantiomer: one of a pair of optical isomers (stereoisomers) that are non-superimposable mirror images. D- and L-glucose are enantiomers.

Endergonic reaction: a chemical reaction in which there is a net input of free energy.

Endothermic reaction: a chemical reaction in which heat energy is absorbed from the surroundings (i.e. $\Delta H > 0$).

Enthalpy: a thermodynamic term that is a measure of the heat content of a system.

Entropy: a thermodynamic term that is a measure of the extent of disorder of a system.

Equilibrium: the position in a chemical reaction where the forward rate of formation of products is equal to the reverse rate of formation of reactants; the reaction is said to be in equilibrium.

Equilibrium constant (K_c, K_{eq}): the equilibrium constant K_c relates to a chemical reaction at equilibrium and can be calculated from a knowledge of the concentration of reactants and products at equilibrium. For enzyme-catalysed reactions the equilibrium constant is usually depicted as K_{eq}. For the reaction $aA + bB \rightleftharpoons cC + dD$, the equilibrium constant is given by the expression:

$$K_c = \frac{[C]^c[D]^d}{[A]^a[B]^b}$$

Ester: an organic molecule that contains an –OR or –OAr group attached to a carbonyl group.

Exergonic reaction: a spontaneous chemical reaction in which there is a net release of free energy.

Exothermic reaction: a chemical reaction in which there is a net release of heat energy (i.e. $DH < 0$).

Facilitated diffusion: passage of a solute along a concentration gradient facilitated by specific proteins (transporters) which speed up the transport process.

Family (of organic compounds): the collection of a large group of organic molecules which have a characteristic functional group and pattern of behaviour in common.

Fatty acid: a long chain carboxylic acid. Fatty acids vary in length and the number and location of double bonds. Double bonds allow for *cis* and *trans* geometric isomers. Three fatty acids linked through a glycerol molecule form a triacylglycerol (fat); two fatty acids linked to glycerol that has a phosphate group form a phospholipid, the major component of biological membranes.

Feasibility: the extent to which (a chemical process) will be successful.

Feedback inhibition: a method of metabolic control in which the end product of a metabolic pathway (or reaction therein) acts as an inhibitor to an enzyme in that pathway.

Filtration: a mechanical operation which is used to separate solids from fluids.

First law of thermodynamics: the principle of conservation of energy. The internal energy of a system is a constant; energy can be neither created nor destroyed.

Free energy: the (Gibbs) free energy is energy that is available to do work. It is related to enthalpy and entropy by the Gibbs equation:

$$DG = DH - TDS$$

Free energy that is released in a catabolic reaction is often coupled to an anabolic reaction, such as ATP synthesis.

Free radical: a molecule or atom with an unpaired electron that has resulted from the homolytic fission of a covalent bond. Radicals are highly reactive.

Functional group: a specific configuration of atoms, often containing a relatively highly electronegative atom, commonly attached to the carbon skeleton of organic molecules, and usually involved in chemical reactions, e.g. –OH, >C=O.

Gas laws: a set of interrelated equations which relate the fundamental properties of gases (pressure, volume, temperature, amount) to each other.

Gay-Lussac's law: the pressure of a fixed amount of gas with a fixed volume is directly proportional to the absolute temperature of the gas.

Geometric isomers: compounds that have the same chemical formula but differ in the spatial arrangement of their atoms. *Cis-trans* isomers are isomers which have different arrangements of atoms about a double bond.

Gibbs free energy: See *Free energy*.

Gluconeogenesis: the synthesis of 'new' glucose from non-sugar precursors such as lactate. Applied more specifically to the synthesis of glucose in the liver.

Glycolysis: the splitting of glucose into pyruvate. Glycolysis is an oxidative metabolic pathway that occurs in all living cells and is an essential provider of cellular energy.

Glycosidic link: a covalent bond formed between two monosaccharides by a dehydration reaction. The most commonly encountered glycosidic bonds are formed between the anomeric carbon of one sugar and a hydroxyl group of another sugar.

Half-life:
i. the time taken for the concentration of a substance to fall by a half (in chemical kinetics)
ii. the time taken for half the original number of radionuclides to disintegrate (in radioactivity).

Half-reaction: an equation which shows the electron loss or gain in an oxidation or reduction.

Henry's law: the solubility of a gas in contact with the surface of a liquid is directly proportional to the partial pressure of the gas.

Henry's law constant (k_H): the constant of proportionality between the partial pressure of a gas above a liquid and the concentration of the gas in the liquid at a fixed temperature.

Heterolytic fission: fission is the process by which a molecule splits into two constituent parts. This occurs when one of the bonds between atoms in the molecule is broken. In heterolytic fission, the two electrons from the broken bond go to the same species. This occurs when one species is significantly more electronegative than the other. Heterolytic fission results in a negatively charged anion, which received both electrons, and a positively charged cation, which received neither.

Homolytic fission: in homolytic fission, the two electrons from the broken bond are shared between the resulting species. This means that each species contains an unpaired electron in an outer shell. They are therefore highly reactive, and are known as free radicals. Homolytic fission occurs when the two atoms being separated have a similar or identical electronegativity; that is, they have roughly the same ability to attract electrons to themselves.

Hybridisation: a model for chemical bonding in which hybrid orbitals are formed.

Hydration shell: a layer of water molecules surrounding a central ion or other species.

Hydrocarbon: an organic molecule consisting of only hydrogen and carbon.

Hydrogen bond: an electrostatic interaction between a hydrogen atom in one molecule and strongly electronegative atom (O, N or F) in a nearby molecule. The interaction may be between two different molecules or within the same molecule.

Hydrogen ion: a hydrogen atom which has lost its single outer electron.

Hydrolysis: the reaction of a substance with water resulting in the cleavage of the compound into two parts, each of which combines with a fragment (H^+ or OH^-) from the water molecule.

Hydrophilic: having an affinity for water. From 'hydro' referring to water and 'philic' meaning 'liking'.

Hydrophobic: having an aversion to water. From 'hydro' referring to water and 'phobic' meaning 'fearing'.

Hydrophobic interaction: a repulsive interaction between a molecule and water; the coalescing of hydrophobic groups caused by their exclusion by water.

Hydroxyl group: a functional group consisting of a hydrogen atom linked to an oxygen atom by a polar covalent bond. Molecules possessing this group are soluble in water and are called alcohols.

Ideal gas: a gas that would obey the gas laws exactly. No known gas is an *ideal gas*. An ideal gas has no intermolecular interactions.

Intermolecular forces: the attractive and repulsive forces which exist between molecules.

Intramolecular forces: the interactions which exist between different fragments of the same molecule.

Ion: an electrically charged atom or group of atoms.

Ionic product of water (K_w): the product of the concentrations of hydroxonium ions and hydroxide ions in an aqueous solution, equal to 1×10^{-14} M^2.

Ionic bond: a chemical bond resulting from the attraction between oppositely charged ions.

Isomer: one of several organic compounds with the same molecular formula but different structures and therefore different properties. The major types of isomerism are structural, geometric and stereoisomers (enantiomers).

Isotope: one of several atomic forms of an element, each containing a different number of neutrons (but the same number of protons) and therefore differing in atomic mass. Isotopes may be stable or unstable (radioisotopes).

Joule: a unit of energy. 1 J = 0.239 cal; 1 cal = 4.184 J.

K_a: See *Acid dissociation constant.*

K_{eq}: See *Equilibrium constant.*

K_w: See *Ionic product of water.*

Ketone: an organic molecule that contains two alkyl groups or two aryl groups or one alkyl and one aryl group attached to a carbonyl group.

Kilocalorie: a unit of energy equal to 4.184 kJ; 1000 calories, or 1 Calorie (see *Calorie*).

Kinetics: the study of the rates and mechanisms of chemical reactions.

Kinetic energy: the energy of a particle due to its motion.

Le Chatelier's principle: states that when a stress is applied to a system at equilibrium, the equilibrium will shift so as to minimise the effects of the stress.

Lone pair: a pair of electrons in the valence shell of an atom, not involved in bonding.

Macromolecule: a large molecule (polymer) formed by the joining together of smaller molecules (monomers) usually by a condensation reaction. Polysaccharides, proteins and nucleic acid are macromolecules.

Metabolic pathway: a connected series of chemical reactions which take place in a living cell.

Metabolism: the totality of an organism's chemical reactions, consisting of catabolic and anabolic pathways.

Molarity: a common measure of solution concentration referring to the number of moles per volume (l or dm^3) of solution.

Molar mass: the mass in grams of one mole of a substance.

Mole: the amount of substance which contains the same number or particles as there are atoms in 12.0 g of the isotope ^{12}C.

Molecule: two or more atoms held together by covalent bonds.

Molecular orbital: a region in space within a molecule that represents the probability of finding an electron.

Monomer: the sub-unit that serves as the building block of a polymer.

Monosaccharide: the simplest carbohydrate (simple sugar), active alone or serving as a monomer in disaccharides or polysaccharides. The molecular formulae of monosaccharides are generally some multiple of CH_2O.

***Mr*:** See *Relative molecular mass.*

Mutarotation: the change in specific rotation that occurs when an α (alpha) or β (beta) hemiacetal form of carbohydrate is converted into an equilibrium mixture of the two forms.

NAD^+ (NADH): nicotinamide adenine dinucleotide is a coenzyme that participates in redox reactions. NAD^+ is the oxidised form that gains two electrons in being reduced to NADH.

Nernst equation: an equation that relates the free energy change in a redox reaction to the change in standard reduction potential ($\Delta E^{o\prime}$) of a reaction.

$$\Delta G^{o\prime} = -nF\Delta E^{o\prime}$$

Neutron: an electrically neutral particle found in the nucleus of an atom. Different numbers of neutrons in the same element give rise to isotopes.

Nucleic acid: a polymer consisting of nucleotide monomers (base–sugar–phosphate). In DNA the sugar is deoxyribose, in RNA the sugar is ribose.

Nucleotide: molecule consisting of a purine or pyrimidine base, a pentose sugar (ribose or deoxyribose) and a phosphate group.

Nucleophile: species that are attracted to positive centres. Typical nucleophiles are negative ions or neutral atoms with a lone pair of electrons.

Nucleus (of an atom): the small positively charged centre of the atom in which most of the mass of the atom is concentrated.

Octet rule: the tendency of atoms to form ionic or covalent bonds in order that their outer (valence) energy level attains a full (8) complement of electrons. Atoms or ions that attain this state are generally stable and unreactive, e.g. the inert gases such as helium and neon.

Optical isomerism: a form of isomerism (specifically stereoisomerism) whereby the two isomers are the same in every way except that they are non-superimposable mirror images of each other. Optical isomers are known as chiral molecules.

Orbital hybridisation: the rearrangement of electrons in the outer energy level of an atom in order to maximise the number of unpaired electrons available to participate in covalent bonding.

Order of reaction: the order of a reaction refers to the number of components involved in the rate-determining step. A first order reaction depends on only one component in the reaction mixture.

Osmosis: the net movement of solvent molecules through a partially permeable membrane into a region of higher solute concentration in order to equalize the concentration.

Osmotic pressure: the pressure which needs to be applied to a solution to prevent the inward flow of water across a partially permeable membrane or the minimum pressure needed to prevent osmosis.

Oxidation: combination with oxygen or loss of electrons by an atom or group of atoms in a reaction.

Oxidation–reduction (redox) reaction: a simultaneous oxidation and reduction reaction.

Oxidative phosphorylation: the synthesis of ATP using energy derived from the redox reactions of an electron transport chain.

Oxidizing agent: a species which causes oxidation by accepting electrons from another species.

Passive transport: entropy-driven transport that does not require chemical energy.

Peptide bond: the covalent secondary amide linkage that joins the carbonyl group of one amino acid to the amino nitrogen of another in peptides and proteins.

Periodic table: an arrangement of the known elements according to their atomic number.

pH: defined as $-\log_{10}[H^+]$ this is a measure of the acidity of a species, solution or system. pH< 7 indicates an acidic solution, pH = 7 indicates a neutral solution, and pH > 7 indicates an alkaline solution.

Phosphodiester: a phosphate group bonded to the rest of the molecule by two ester groups.

pK_a: a logarithmic value that indicates the strength of an acid. pK_a is defined as the negative logarithm of the acid dissociation constant K_a.

Photon: a quantum of light energy.

Phospholipid: a molecule consisting of a polar hydrophilic glycerol + phosphate head-group, and a non-polar hydrophobic 'tail' consisting of two fatty acids. The molecule is amphipathic. Phospholipids are major constituent of biological membranes.

Pi bonds: the covalent bond formed between two atoms that have an unpaired electron in a p orbital, and formed by the 'side-to-side' merging of the p orbitals, above and below the bond axis. Always found in addition to a sigma covalent bond between the two atoms, hence the description of a 'double bond'. Double bonds are considered as a functional group since they constitute a region of increased electron density that may be 'attacked' by an electrophilic reagent.

Polar: a term used to describe species which have an uneven distribution of charge – often indicated by δ+/δ–.

Polar molecule (group): a molecule that possesses an electrical dipole; for example, those functional groups containing a relatively highly electronegative atom are said to be polar. Polar molecules (groups) are generally soluble in water. Apolar molecules do not possess such groups and are not soluble in water.

Polar covalent bond: a covalent bond in which the two atoms differ in their electronegativities. The more electronegative atom draws electrons towards itself becoming slightly negatively charged, with the other atom becoming slightly positively charged.

Polymer: a large molecule formed from many monomers.

Potential energy: the energy an object has because of its position, rather than its motion. The bond energy between two atoms is potential energy until that bond is broken.

Pressure (of a gas): the force a gas exerts on the sides of a container by virtue of the collisions of molecules of the gas with the container

Protein: a large polymer constructed from more than 50 a-amino acid units (monomers) linked through peptide bonds.

Racemic mixture: an equimolar mixture of enantiomers of a molecule, e.g. a mixture of D- and L-glucose.

Radical: a molecule or atom with an unpaired electron that has resulted from the homolytic fission of a covalent bond. Radicals are highly reactive.

Radioactivity: the emission of α or β particles or g radiation by an atom (or a combination thereof).

Radioisotope: an unstable form of an element in which the nucleus decays spontaneously emitting particles (α or β particles) and/or energy (γ radiation).

Rate constant: the constant of proportionality in a rate equation.

Rate equation: an expression of the observed relationship between the rate of a reaction and the concentration of each reactant.

Rate-limiting step: a number of intermediate states may be involved in the conversion of a reactant to a product. The intermediate that is formed at the slowest rate constitutes the rate-limiting step for the overall reaction.

Rate of reaction: the rate of any chemical reaction is defined as the change in concentration of a component of the reaction over a period of time.

Reactant: a substance that participates in a chemical reaction.

Reaction mechanism: the reaction steps and the nature of the intermediates involved in the conversion of reactant to product.

Reaction order: See *Order of reaction.*

Reactive site: a site on a molecule, in or on a functional group, which is either electron deficient or electron rich, that will attract attack by a nucleophilic or electrophilic agent respectively.

Redox reaction: two coupled half-reactions, one an oxidation and the other a reduction.

Redox potential: a measure of an atom's tendency to gain or lose electrons. Atoms with a high redox potential (a positive value) gain electrons from atoms with a lower redox potential (a more negative value).

Reducing agent: a substance which causes reduction by donating electrons.

Reduction: the gain of electrons by a substance.

Reduction potential (E): a measure of the tendency of a substance to reduce other substances. The more negative the reduction potential, the greater the tendency to donate electrons.

Relative molecular mass (M_r): the mass of a molecule relative to one-twelfth the mass of ^{12}C.

Salt:

i. an ionic compound that can be formed by replacing one or more of the hydrogen ions of an acid with another positive ion.
ii. the product, formed along with water, from the reaction of an acid with a base.

Saturated: referring to a biological molecule that does not contain a carbon–carbon double bond. Also: unable to take up further material; can refer to compounds which can take up no further hydrogen or solutions which can take no further solute.

Second law of thermodynamics: the principle whereby energy transfers or transformations increase the disorder of the universe. A spontaneous change is always accompanied by an increase in entropy of the system. Ordered forms of energy (in covalent bonds) are at least partly converted to heat energy during metabolic reactions, leading to an increase in disorder (an increase in entropy) of the surroundings.

Sigma bond: a covalent bond formed from the overlap of atomic orbitals along the bond axis ('head-to-head'), as opposed to pi bonds where the orbital overlap is below and above the bond axis.

Solute: a substance that is dissolved in a solution.

Solution: a homogeneous liquid mixture of two or more substances.

Solvent: the dissolving agent of a solution. Water is the most versatile solvent.

Spontaneous change: a change which occurs naturally and does not need to be driven.

Standard state (standard free energy change, $\Delta G^{\circ\prime}$) (standard reduction potential, $\Delta E^{\circ\prime}$): A set of reference conditions for a chemical reaction. In biochemistry the standard state is defined as that at a temperature of 298 K (25°C), a pressure of one atmosphere, a solute concentration of 1 M, and a pH of 7.0.

Stereoisomer: a molecule that is a mirror image of another molecule with the same molecular formula; these isomers have atoms with the same connectivity but are arranged differently in space.

Stoichiometry: the ratio of number of moles of reactants consumed to products formed in the chemical equation for a reaction.

Structural formula: a molecular notation in which the individual atoms of a molecule are shown joined together by lines representing covalent bonds.

Structural isomers: compounds that have the same molecular formula but differ in the covalent arrangement of their atoms.

Substitution reaction: a reaction, normally involving organic compounds, whereby an atom, or group of atoms, is replaced by another atom, or group of atoms.

Substrate: the reactant on which an enzyme works.

Sulphydryl group: a functional group consisting of a sulphur atom bonded to a hydrogen atom.

Tetrahedral: the geometry adopted around a saturated carbon atom; four atoms (or groups) covalently bonded to one carbon atom will tend to point towards the corners of a tetrahedron.

Transition state: an unstable high energy arrangement of atoms in which chemical bonds are being formed or broken. Transition states have structures between those of the substrate (reactant) and product of the reaction.

Universal gas constant: a physical constant (R) that relates the pressure, volume, number of moles and absolute temperature of a gas according to the following equation: $pv = nRT$. Has a value of 8.3145 joules per kelvin per mol.

Unpaired electron: a single electron in an outer (valence) energy level that may participate in covalent bond formation.

Unsaturated fatty acid: a fatty acid with at least one carbon–carbon double bond. In general the double bonds of an unsaturated fatty acid are of the *cis* configuration.

Valence: the bonding capacity of an atom; generally equal to the number of unpaired electrons in an atom's outer electron energy level.

Valence electron: an electron in a valence shell which can undergo bonding with other valence electrons.

Valence shell: the outermost energy shell (energy level) of an atom containing the valence electrons involved in the chemical reactions of that element.

van der Waals forces: weak charge–charge attractions or repulsions between molecules (intermolecular) or parts of molecules in close proximity. Such forces are the result of temporary transient dipoles on molecules leading to localised charge fluctuations.

Index